W9-AYE-911

HEREDITY
The Code of Life

Anthea Maton
Former NSTA National Coordinator
Project Scope, Sequence, Coordination
Washington, DC

Jean Hopkins
Science Instructor and Department Chairperson
John H. Wood Middle School
San Antonio, Texas

Susan Johnson
Professor of Biology
Ball State University
Muncie, Indiana

David LaHart
Senior Instructor
Florida Solar Energy Center
Cape Canaveral, Florida

Maryanna Quon Warner
Science Instructor
Del Dios Middle School
Escondido, California

Jill D. Wright
Professor of Science Education
Director of International Field Programs
University of Pittsburgh
Pittsburgh, Pennsylvania

Prentice Hall
Englewood Cliffs, New Jersey
Needham, Massachusetts

Prentice Hall Science

Heredity: The Code of Life

Student Text and Annotated Teacher's Edition
Laboratory Manual
Teacher's Resource Package
Teacher's Desk Reference
Computer Test Bank
Teaching Transparencies
Product Testing Activities
Computer Courseware
Video and Interactive Video

The illustration on the cover, rendered by Joseph Cellini, can be used to demonstrate the concept of heredity and variation within a species.

Credits begin on page 123.

SECOND EDITION

ISBN 0-13-400490-6

12 13 14 15 99 98

Prentice Hall
A Division of Simon & Schuster
Englewood Cliffs, New Jersey 07632

STAFF CREDITS

Editorial:	Harry Bakalian, Pamela E. Hirschfeld, Maureen Grassi, Robert P. Letendre, Elisa Mui Eiger, Lorraine Smith-Phelan, Christine A. Caputo
Design:	AnnMarie Roselli, Carmela Pereira, Susan Walrath, Leslie Osher, Art Soares
Production:	Suse F. Bell, Joan McCulley, Elizabeth Torjussen, Christina Burghard
Photo Research:	Libby Forsyth, Emily Rose, Martha Conway
Publishing Technology:	Andrew Grey Bommarito, Deborah Jones, Monduane Harris, Michael Colucci, Gregory Myers, Cleasta Wilburn
Marketing:	Andrew Socha, Victoria Willows
Pre-Press Production:	Laura Sanderson, Kathryn Dix, Denise Herckenrath
Manufacturing:	Rhett Conklin, Gertrude Szyferblatt

Consultants

Kathy French	National Science Consultant
Jeannie Dennard	National Science Consultant
Brenda Underwood	National Science Consultant
Janelle Conarton	National Science Consultant

Contributing Writers

Linda Densman
Science Instructor
Hurst, TX

Linda Grant
Former Science Instructor
Weatherford, TX

Heather Hirschfeld
Science Writer
Durham, NC

Marcia Mungenast
Science Writer
Upper Montclair, NJ

Michael Ross
Science Writer
New York City, NY

Content Reviewers

Dan Anthony
Science Mentor
Rialto, CA

John Barrow
Science Instructor
Pomona, CA

Leslie Bettencourt
Science Instructor
Harrisville, RI

Carol Bishop
Science Instructor
Palm Desert, CA

Dan Bohan
Science Instructor
Palm Desert, CA

Steve M. Carlson
Science Instructor
Milwaukie, OR

Larry Flammer
Science Instructor
San Jose, CA

Steve Ferguson
Science Instructor
Lee's Summit, MO

Robin Lee Harris Freedman
Science Instructor
Fort Bragg, CA

Edith H. Gladden
Former Science Instructor
Philadelphia, PA

Vernita Marie Graves
Science Instructor
Tenafly, NJ

Jack Grube
Science Instructor
San Jose, CA

Emiel Hamberlin
Science Instructor
Chicago, IL

Dwight Kertzman
Science Instructor
Tulsa, OK

Judy Kirschbaum
Science/Computer Instructor
Tenafly, NJ

Kenneth L. Krause
Science Instructor
Milwaukie, OR

Ernest W. Kuehl, Jr.
Science Instructor
Bayside, NY

Mary Grace Lopez
Science Instructor
Corpus Christi, TX

Warren Maggard
Science Instructor
PeWee Valley, KY

Della M. McCaughan
Science Instructor
Biloxi, MS

Stanley J. Mulak
Former Science Instructor
Jensen Beach, FL

Richard Myers
Science Instructor
Portland, OR

Carol Nathanson
Science Mentor
Riverside, CA

Sylvia Neivert
Former Science Instructor
San Diego, CA

Jarvis VNC Pahl
Science Instructor
Rialto, CA

Arlene Sackman
Science Instructor
Tulare, CA

Christine Schumacher
Science Instructor
Pikesville, MD

Suzanne Steinke
Science Instructor
Towson, MD

Len Svinth
Science Instructor/
Chairperson
Petaluma, CA

Elaine M. Tadros
Science Instructor
Palm Desert, CA

Joyce K. Walsh
Science Instructor
Midlothian, VA

Steve Weinberg
Science Instructor
West Hartford, CT

Charlene West, PhD
Director of Curriculum
Rialto, CA

John Westwater
Science Instructor
Medford, MA

Glenna Wilkoff
Science Instructor
Chesterfield, OH

Edee Norman Wiziecki
Science Instructor
Urbana, IL

Teacher Advisory Panel

Beverly Brown
Science Instructor
Livonia, MI

James Burg
Science Instructor
Cincinnati, OH

Karen M. Cannon
Science Instructor
San Diego, CA

John Eby
Science Instructor
Richmond, CA

Elsie M. Jones
Science Instructor
Marietta, GA

Michael Pierre McKereghan
Science Instructor
Denver, CO

Donald C. Pace, Sr.
Science Instructor
Reisterstown, MD

Carlos Francisco Sainz
Science Instructor
National City, CA

William Reed
Science Instructor
Indianapolis, IN

Multicultural Consultant

Steven J. Rakow
Associate Professor
University of Houston—Clear Lake
Houston, TX

English as a Second Language (ESL) Consultants

Jaime Morales
Bilingual Coordinator
Huntington Park, CA

Pat Hollis Smith
Former ESL Instructor
Beaumont, TX

Reading Consultant

Larry Swinburne
Director
Swinburne Readability Laboratory

CONTENTS

HEREDITY: THE CODE OF LIFE

Activity Bank/Reference Section

Features

CONCEPT MAPPING

Throughout your study of science, you will learn a variety of terms, facts, figures, and concepts. Each new topic you encounter will provide its own collection of words and ideas—which, at times, you may think seem endless. But each of the ideas within a particular topic is related in some way to the others. No concept in science is isolated. Thus it will help you to understand the topic if you see the whole picture; that is, the interconnectedness of all the individual terms and ideas. This is a much more effective and satisfying way of learning than memorizing separate facts.

Actually, this should be a rather familiar process for you. Although you may not think about it in this way, you analyze many of the elements in your daily life by looking for relationships or connections. For example, when you look at a collection of flowers, you may divide them into groups: roses, carnations, and daisies. You may then associate colors with these flowers: red, pink, and white. The general topic is flowers. The subtopic is types of flowers. And the colors are specific terms that describe flowers. A topic makes more sense and is more easily understood if you understand how it is broken down into individual ideas and how these ideas are related to one another and to the entire topic.

It is often helpful to organize information visually so that you can see how it all fits together. One technique for describing related ideas is called a **concept map**. In a concept map, an idea is represented by a word or phrase enclosed in a box. There are several ideas in any concept map. A connection between two ideas is made with a line. A word or two that describes the connection is written on or near the line. The general topic is located at the top of the map. That topic is then broken down into subtopics, or more specific ideas, by branching lines. The most specific topics are located at the bottom of the map.

To construct a concept map, first identify the important ideas or key terms in the chapter or section. Do not try to include too much information. Use your judgment as to what is really important. Write the general topic at the top of your map. Let's use an example to help illustrate this process. Suppose you decide that the key terms in a section you are reading are School, Living Things, Language Arts, Subtraction, Grammar, Mathematics, Experiments, Papers, Science, Addition, Novels. The general topic is School. Write and enclose this word in a box at the top of your map.

SCHOOL

Now choose the subtopics—Language Arts, Science, Mathematics. Figure out how they are related to the topic. Add these words to your map. Continue this procedure until you have included all the important ideas and terms. Then use lines to make the appropriate connections between ideas and terms. Don't forget to write a word or two on or near the connecting line to describe the nature of the connection.

Do not be concerned if you have to redraw your map (perhaps several times!) before you show all the important connections clearly. If, for example, you write papers for Science as well as for Language Arts, you may want to place these two subjects next to each other so that the lines do not overlap.

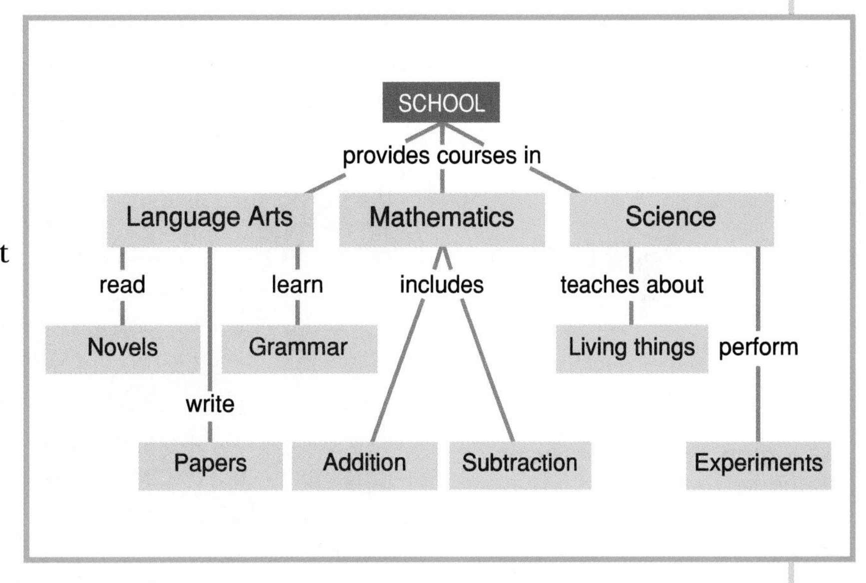

One more thing you should know about concept mapping: Concepts can be correctly mapped in many different ways. In fact, it is unlikely that any two people will draw identical concept maps for a complex topic. Thus there is no one correct concept map for any topic! Even though your concept map may not match those of your classmates, it will be correct as long as it shows the most important concepts and the clear relationships among them. Your concept map will also be correct if it has meaning to you and if it helps you understand the material you are reading. A concept map should be so clear that if some of the terms are erased, the missing terms could easily be filled in by following the logic of the concept map.

HEREDITY

The Code of Life

A unicorn is a mythical animal with the body and head of a horse, the hind legs of a stag, the tail of a lion, and a single horn in the middle of its forehead. The unicorn was not the only mythical beast with a combination of body parts of different animals. The chimera of Greek mythology had the head of a lion, the body of a goat, and the tail of a serpent. In

The unicorn is a mythical animal that could not exist in real life. However, many other plants and animals almost as bizarre as the unicorn do live on Earth.

This young horse has been bred to have the same long legs and strong muscles as its mother. Will it grow up to be a champion racehorse?

Notice the family resemblance inherited by the children from their parents.

CHAPTERS

ancient Egypt, the Sphinx was constructed as a winged lion with a woman's head. Do you think any of these fabulous creatures could exist in the real world?

Unfortunately, the unicorn, chimera, and Sphinx exist only in people's imaginations. In the real world, living things always resemble their parents. Horses never give birth to unicorns. In this book, you will meet the man who discovered how living things pass their characteristics on to their offspring. You will explore the complex molecule—called DNA—that makes heredity possible and also see how the principles of heredity that apply to plants and animals apply to humans as well. Finally, you will learn how scientists are beginning to use genetic engineering to produce organisms that will benefit humans in many ways—organisms almost as exotic, in their own way, as the creatures of myth.

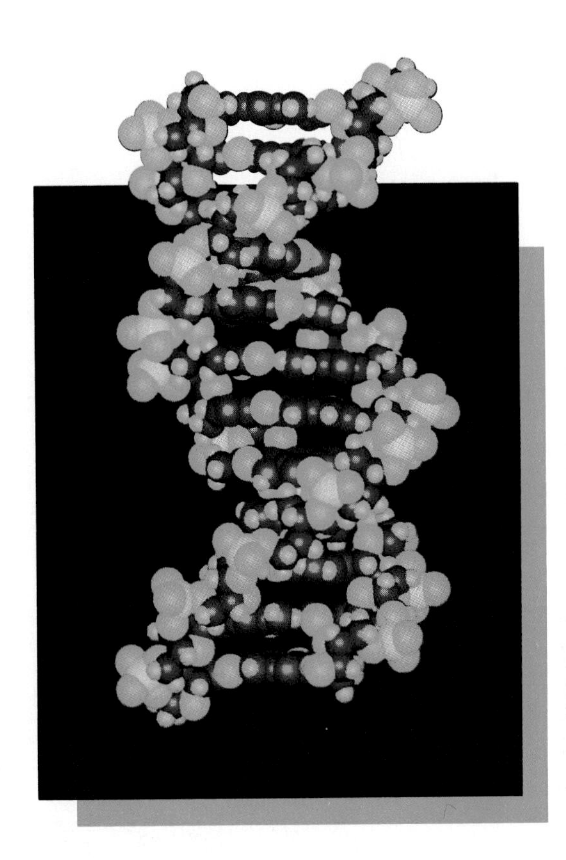

A computer-generated model of DNA is shown in this photograph.

Discovery Activity

Variations on a Theme

Think of a particular human trait, or characteristic—such as hair color, eye color, or skin color. Look around your classroom. How many different variations of this trait do you see among your classmates? For example, how many have red hair? How many have brown hair? Record your observations in a chart or table.

- Based on your observations, which form of the trait is most common in your classroom? Least common? Do you think there is any reason for this?

What Is Genetics?

Guide for Reading

After you read the following sections, you will be able to

1–1 History of Genetics

- Describe how traits are passed from one generation to another.
- Explain the difference between dominant and recessive traits.

1–2 Principles of Genetics

- State the law of segregation and the law of independent assortment.
- Explain what is meant by incomplete dominance.

1–3 Genetics and Probability

- Relate the law of probability to the study of genetics.
- Describe how Punnett squares can be used to predict the results of genetic crosses.

Do you have a cat, or do you have a friend who has a cat? You probably know that there are many different kinds of cats. Some cats have long, soft hair and fluffy tails. Others have short, curly hair and skinny tails. There is even one kind of cat that appears to have no hair at all! Despite these, and many other, variations, you can still recognize these animals as cats. Why is this so? What makes one cat different from another and yet still recognizable as a member of the cat family? The answer can be found in the science of genetics.

The study of genetics explains why one cat is different from all other cats and also why a cat is different from a human or a maple tree. All living things, including cats, resemble their parents. But each individual also has certain unique characteristics that make it different from every other living thing on Earth. As you will learn in this chapter, the history of genetics is a fascinating story of mystery and discovery. The story begins with one man in a garden more than 100 years ago. . . .

Journal *Activity*

You and Your World Have you ever seen a litter of kittens or puppies? Did all the kittens or puppies look exactly alike? In your journal, describe how they were alike and how they were different. Do you have any idea why they look as they do?

These kittens may have different colors and markings, but they all look like cats.

Guide for Reading

Focus on this question as you read.

- *What is meant by the term genetics?*

1–1 History of Genetics

The history of genetics began with a monk named Gregor Mendel working in the garden of a small monastery in eastern Europe. Mendel, whose parents were Austrian peasants, was born in 1822. He entered the monastery at the age of 21 and was ordained a priest 4 years later. In 1851, Mendel was sent to the University of Vienna to study science and mathematics. After he left the university, Mendel spent the next 14 years working at the monastery and teaching at a nearby high school. In addition to teaching, Mendel also looked after the monastery garden. Here he grew hundreds of pea plants. Mendel experimented with the pea plants to see if he could find a pattern in the way certain characteristics were handed down from one generation of pea plants to the next.

Mendel chose pea plants for his experiments for several reasons. Pea plants grow and reproduce quickly. So he knew that he could study many generations of pea plants in a short time. Mendel also knew that pea plants had a variety of different characteristics, or **traits,** that could be studied at the same time. Pea plant traits include how tall the plants grow, the color of their seeds, and the shape of their seeds. Mendel could study all of these traits (as well as other traits) in the same experiment. In addition, pea plants could be crossed, or bred, easily.

Figure 1–1 *Gregor Mendel is shown in his garden studying how traits are passed on from parents to offspring. What organisms did Mendel study?*

The Work of Gregor Mendel

Figure 1–2 shows what the flowers of Mendel's pea plants look like. As in most flowering plants, the flowers of pea plants contain stamens, or male reproductive structures. Stamens produce pollen, which contains male sex cells, or sperm cells. The flowers also contain the female reproductive structure, called the pistil. The pistil produces the female sex cell, or egg cell. When pollen lands on top of the pistil of a flower, pollination occurs. Pollination produces seeds for the next generation of pea plants.

Usually, a pea plant pollinates itself. This type of pollination is known as self-pollination. In self-pollination, pollen from the stamen of one flower lands

Figure 1–2 *Stamens produce pollen, which contains sperm cells. The pistil produces eggs. As a result of pollination, fertilized eggs develop into seeds. When planted, a seed grows into a new pea plant.*

on the pistil of the same flower or on the pistil of a different flower on the same plant. But Mendel found that he could transfer pollen from the stamen of one flower to the pistil of another flower on a different plant. This type of pollination is known as cross-pollination. By using cross-pollination, Mendel was able to cross pea plants with different traits.

Although Mendel did not realize it at the time, his experiments would come to be considered the

Activity Bank

Tulips Are Better Than One, p.104

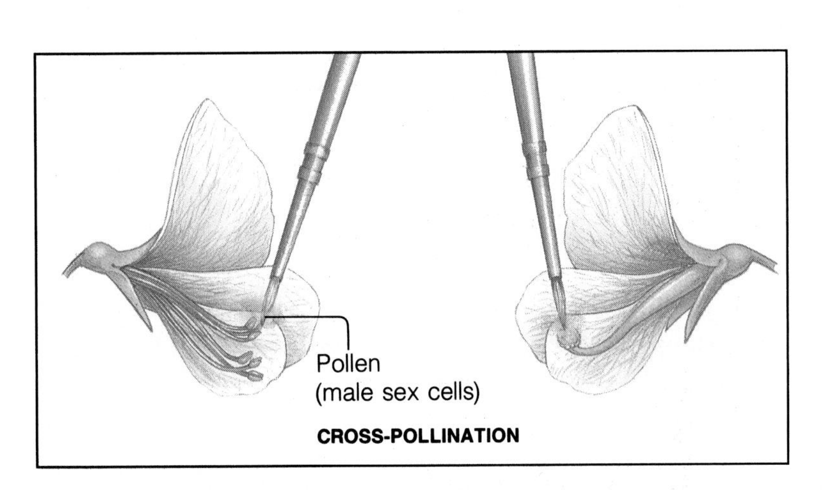

Figure 1–3 *The pollen from one flower is transferred to the pistil of another flower on a different pea plant. What is this process called?*

Genetic Scientists

You may be interested in finding out more about some of the people listed below. They are all scientists who have contributed to the study of genetics and heredity. The library has many resources to help you find out about these people and their work. Choose one of these people and find out what she or he discovered about heredity and genetics. Present your findings in a report.

Rosalind Franklin
James D. Watson
Martha Chase
Jacques Monod
Barbara McClintock

beginning of the science of **genetics** (juh-NEHT-ihks). For this reason, Mendel is called the Father of Genetics. **Genetics is the study of heredity, or the passing on of traits from an organism to its offspring.**

Mendel's Experiments

Mendel began his experiments by first crossing two short pea plants (pea plants with short stems). He discovered that when he planted the seeds from these pea plants with short stems, only short-stemmed plants grew. In other words, members of the next generation of short-stemmed plants were also short-stemmed. This result was what he, and everyone else at that time, expected. New generations of plants always resembled the parent plants. Mendel called these short plants true-breeding plants. By true-breeding plants, Mendel meant those plants that always produce offspring with the same traits as the parents.

In the experiments that followed, Mendel tried crossing two tall pea plants (pea plants with long stems). He wondered if the tall pea plants would also be true-breeding. To his surprise, he found that tall pea plants would not always be true-breeding. Some tall pea plants produced all tall plants. However, other tall pea plants produced mostly tall and some short pea plants. This result was different from the cross between the short pea plants, which produced only short plants. Although he could not explain his results at the time, Mendel realized that there must

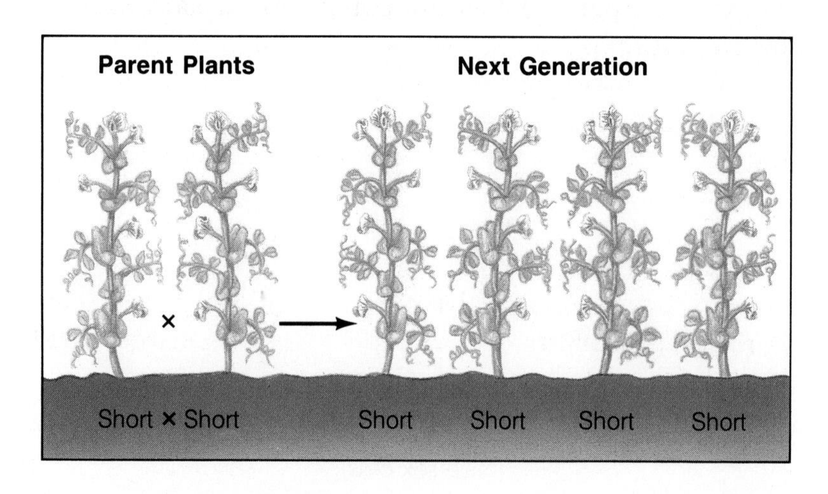

Figure 1–4 *In Mendel's first experiment with pea plants, crossing two short plants resulted in offspring that were all short as well. What did Mendel call these short plants?*

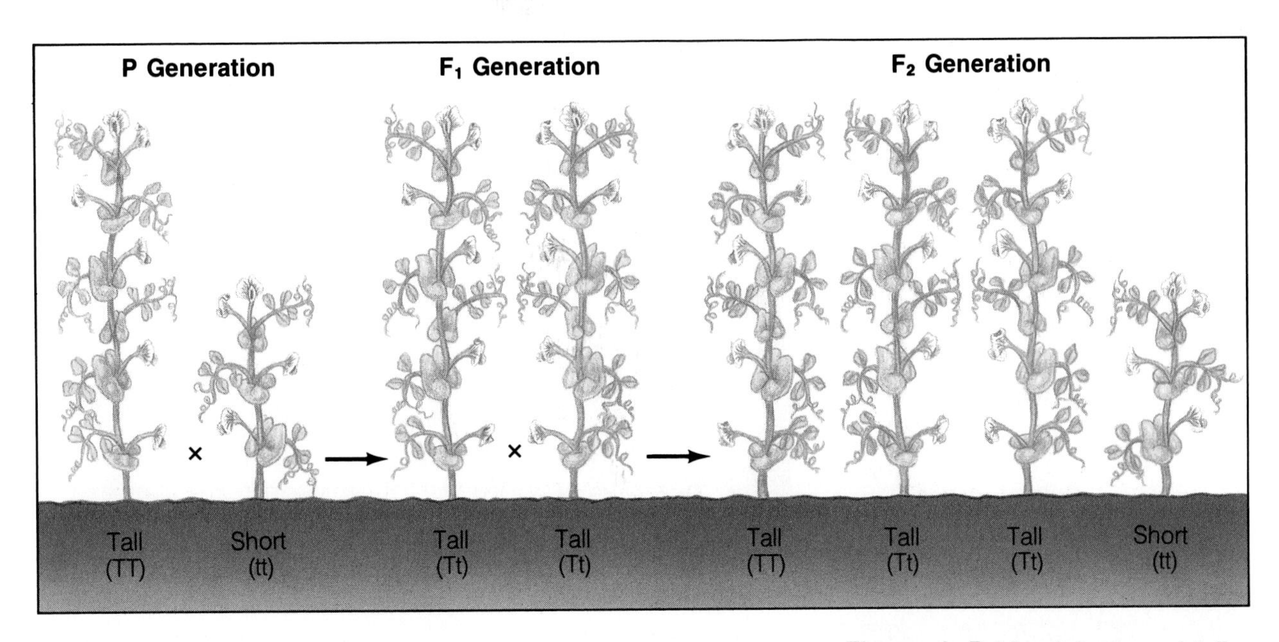

Figure 1–5 *Mendel crossed tall and short pea plants. He discovered that the offspring in the first generation were all tall. What kind of plants were produced in the second generation?*

be two kinds of tall pea plants: true-breeding plants and plants that did not breed true.

Mendel then wondered what would happen if he took pollen from a plant that produced only tall plants (a true-breeding plant) and dusted it onto the pistil of a short plant (another true-breeding plant). To identify the different generations of plants, Mendel gave them different names. He called the first two parent plants the parental generation, or P generation. He called the offspring of the P generation the first filial (FIHL-ee-uhl) generation, or F_1 generation. (The word filial comes from the Latin word *filius,* which means son.) Mendel discovered that all of the plants in the F_1 generation were tall. There were no short plants at all! It was as if the trait for shortness from one of the parent plants had disappeared completely. Mendel could not explain these results either.

What happened next was even more of a mystery. Mendel covered the tall plants of the F_1 generation and allowed them to self-pollinate. That is, the pollen of a flower was allowed to fall onto the pistil of the same flower. Mendel expected that the tall plants would again produce only tall plants. But once again he was surprised. Mendel discovered that some of the plants in what he called the second filial generation, or F_2 generation, were tall and some were short. The trait for shortness seemed to have reappeared! How could this have happened?

ACTIVITY

DISCOVERING

How Do You Measure Up?

1. Choose a partner. Measure each other's height to the nearest centimeter. Record your measurements.

2. On the chalkboard, compile a list showing the height of each student in your class.

3. Make a bar graph showing the number of students for each recorded height measurement.

■ Based on your graph, do you think height in humans is determined in the same way as stem length in pea plants? Give reasons for your answer.

Figure 1–6 *To Mendel's surprise, when he crossed two tall pea plants from the F_1 generation, the trait of shortness reappeared in the F_2 generation. Why did the shortness trait reappear?*

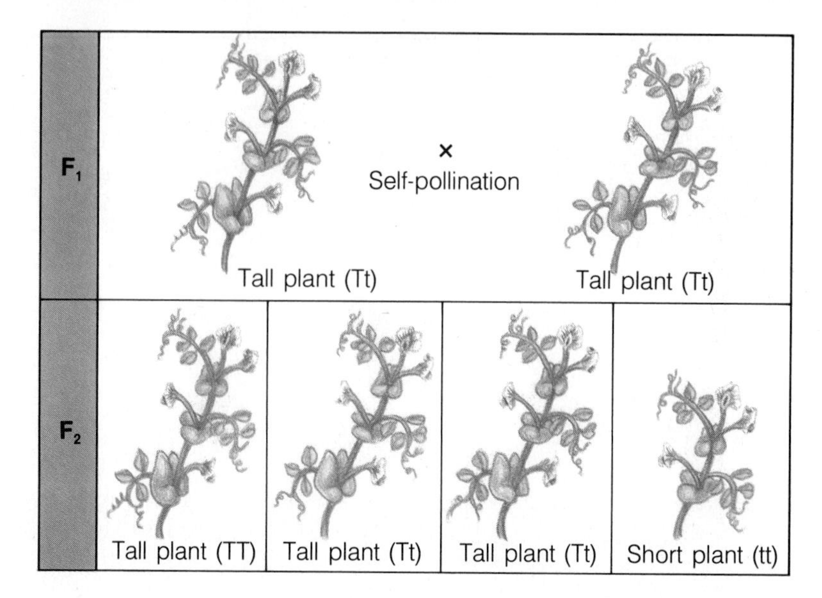

ACTIVITY
DISCOVERING

Observing Traits

Visit a garden center or greenhouse. Take a notebook and a pencil to record your observations.

1. Choose one type of flowering plant to observe, such as petunias, marigolds, or chrysanthemums.

2. To observe the genetic traits of the flowering plant you chose, you must observe 10 of these plants. Look closely at each of the plants.

3. Note common and uncommon traits among the plants, such as the shape of the leaves or the color of the flower petals.

What common traits did you observe on most of the plants?

What uncommon traits did you see on one or more of the plants but not on most of them?

■ How can you determine which traits are dominant?

From the careful records he kept of all his experiments, Mendel made several important discoveries. He observed that the tall plants of the F_1 generation did not breed true. So he reasoned that these plants had to contain factors for both tallness and shortness. When both factors were present in a plant, only tallness showed. These factors, which Mendel called "characters," are now called **genes.** Genes are the units of heredity.

Dominant and Recessive Traits

From his observations, Mendel also concluded that when he crossed two true-breeding plants with opposite traits (tallness and shortness, for example), the offspring plants showed only one of the traits (tallness). That trait seemed to be "stronger" than the other trait (shortness). The stronger trait is called the **dominant** trait. The "weaker" trait, or the trait that seemed to disappear, is called the **recessive** trait.

Geneticists—scientists who study heredity—use symbols to represent the different forms of a gene. A dominant form is represented by a capital letter. For example, the gene form for tallness in pea plants is T. A recessive form is represented by a small, or lowercase, letter. Thus, shortness is t. Every organism has two forms of the gene for each trait. So the symbol for a true-breeding tall plant is TT. The symbol for a true-breeding short plant is tt.

Figure 1–7 *All organisms, even these rabbits, show a combination of dominant and recessive traits. How would you describe the traits of each rabbit?*

1–1 Section Review

1. What is genetics?
2. Compare dominant and recessive traits.

Critical Thinking—*Making Inferences*

3. When Mendel crossed pea plants that produced only round seeds with plants that produced only wrinkled seeds, all the plants in the F_1 generation produced round seeds. However, in the F_2 generation some plants produced wrinkled seeds. Which trait—round seeds or wrinkled seeds—is dominant? Which is recessive? Explain your answers.

ACTIVITY DOING

Dominant and Recessive Traits

1. Obtain two coins.

2. Cut four small, equal-sized pieces of masking tape to fit on the coins without overlapping the edges.

3. Place a piece of tape on each side of both coins.

4. Write a capital letter T on one side of each coin and a lowercase letter t on the other side.

5. Toss both coins together 100 times. Record the letters of the genetic makeup for each toss of the coins.

What are the possible gene combinations? What is the percentage of each?

CONNECTIONS

Breeding the Purr-fect Cat

Did you know that cats are now more popular than dogs as pets in the United States? Many people adopt homeless cats from animal shelters, and others prefer to buy purebred cats from cat breeders. Two of the most popular kinds of purebred cats are Siamese cats and Persian cats. Both of these breeds are very old. Siamese cats have long, thin bodies. They are short-haired, with dark markings on the face, feet, and tail. Persian cats are long-haired, with short, stocky bodies.

In the 1930s, these two breeds were combined in a series of genetics experiments carried out by scientists at Harvard Medical School. The scientists were trying to find out how certain traits in cats are inherited. The result of their work was a new, artificial breed of cat called the Himalayan. The first Himalayan cat was born in 1935. The kitten had long hair like a Persian and the markings of a Siamese.

After the birth of the first Himalayan kitten, professional cat breeders took over. In the 1960s, Himalayans were recognized by groups, such as the Cat Fanciers Association (CFA), that sponsor major cat shows in the United States. Owners of Himalayan cats can thank Gregor Mendel for their pets. Without the science of genetics, which Mendel helped establish, these beautiful cats might never have existed.

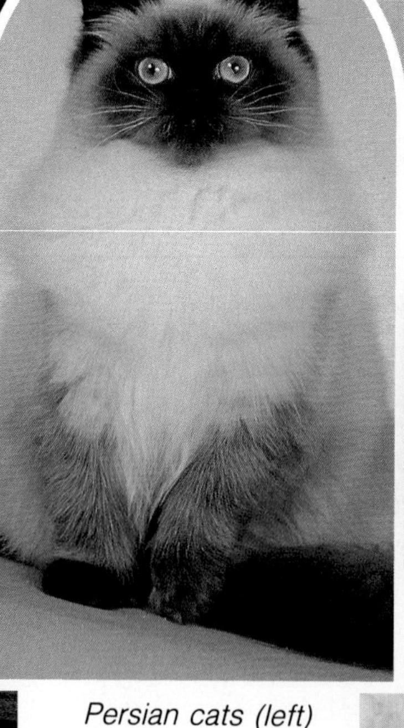

Persian cats (left)
Himalayan cat (center)
Siamese cat (right)

1–2 Principles of Genetics

Guide for Reading

Focus on this question as you read.

- *What are six basic principles of genetics?*

As you read in the previous section, one of the reasons Mendel chose pea plants for his experiments was that they showed a variety of different traits that could be studied at the same time. So in addition to height, or stem length, Mendel also studied seed shape, seed color, seed coat color, pod shape, pod color, and flower position. These traits are illustrated in Figure 1–8. For every trait studied, the results were always the same: Crossing two true-breeding plants with opposite traits did not result in a mixture of the traits. Only one of the traits—the dominant one—appeared in the offspring. But in the next generation, the trait that seemed to disappear—the recessive one—reappeared.

As an example, let's examine a cross between a plant with yellow seeds (YY) and a plant with green seeds (yy). The seeds that are produced in the F_1 generation are all yellow, not a mixture of green and yellow. Why? The gene for yellow seeds, Y, is dominant. It masks, or hides, the recessive gene for green seeds, y. Therefore, all the seeds are yellow. What happens when a plant from the F_1 generation pollinates itself? In this case, most of the seeds are

Figure 1–8 *The chart shows the seven characteristics that Mendel studied in pea plants. Each characteristic has a dominant gene and a recessive gene. Which seed color is dominant in pea plants?*

PEA PLANT TRAITS

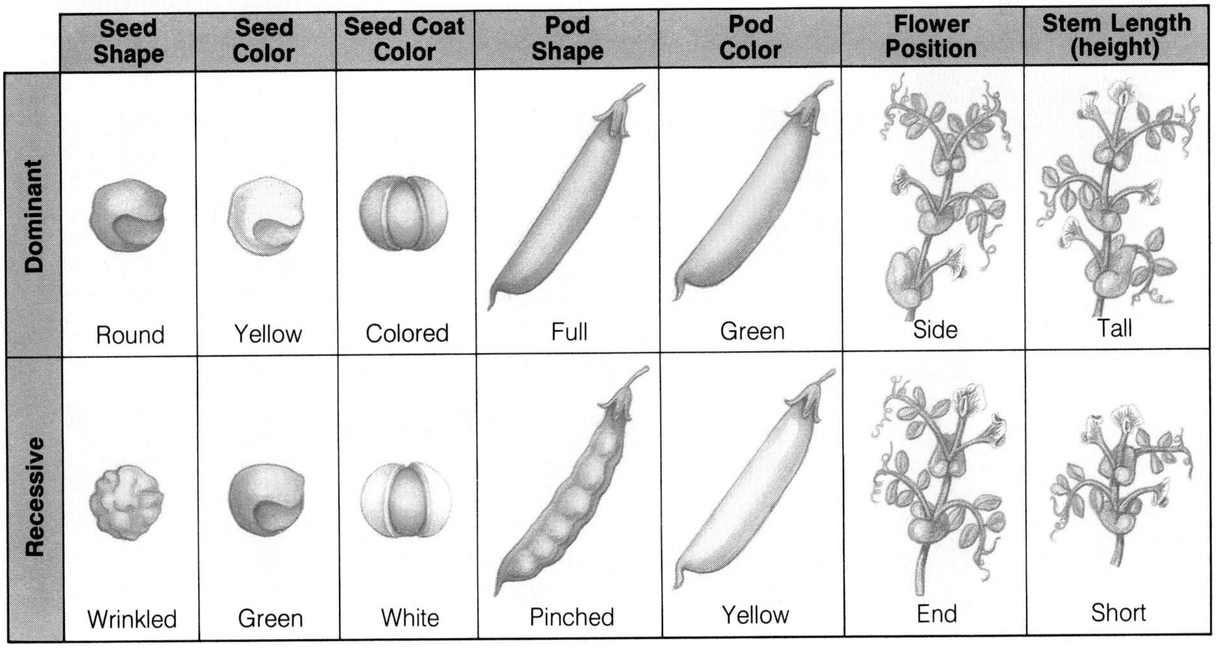

	Seed Shape	Seed Color	Seed Coat Color	Pod Shape	Pod Color	Flower Position	Stem Length (height)
Dominant	Round	Yellow	Colored	Full	Green	Side	Tall
Recessive	Wrinkled	Green	White	Pinched	Yellow	End	Short

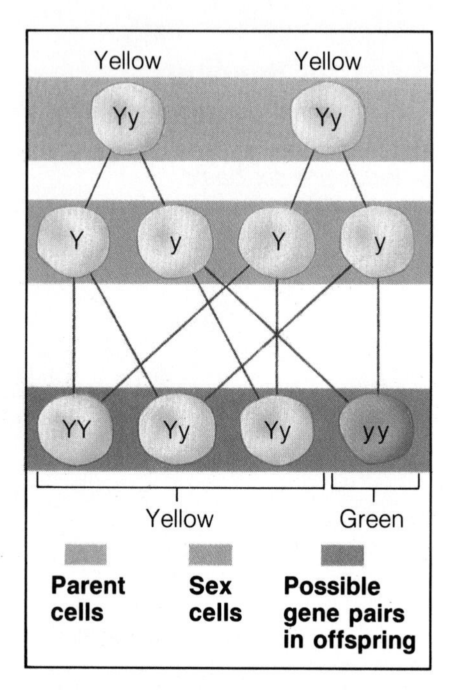

Figure 1–9 *Mendel discovered that a pea plant with green seeds can develop from a cross between parents with yellow seeds. How does this happen?*

yellow, but some are green. The recessive gene for green seeds reappears in the F_2 generation.

An organism that has genes that are alike for a particular trait, such as YY or yy, is called a purebred. An organism that has genes that are different for a trait, such as Yy, is called a **hybrid** (HIGH-brihd). The plants with yellow seeds in Mendel's F_1 generation were hybrid plants (Yy). They were produced by crossing two purebred plants with opposite traits (YY and yy). Many of the plants advertised in seed catalogs are hybrids that were developed by plant breeders. You will learn more about plant and animal hybrids in Chapter 4.

Laws of Genetics

After performing hundreds of experiments and analyzing his observations, Mendel formed a hypothesis about how traits were passed on from one generation of pea plants to another. (Remember, a hypothesis is a suggested explanation for a scientific problem.) Mendel's hypothesis was that each pea plant had a pair of factors, or genes, for each trait. Each parent pea plant could contribute only one gene of each pair to each plant in the next generation. In that way, each plant in the next generation also had a pair of genes for each trait, one from each parent.

Now Mendel could account for the fact that a pea plant with green seeds can develop from a cross

Figure 1–10 *These colorful zinnias (left) and lovely purple petunias and rose-colored vincas (right) are hybrids that were produced by plant breeders.*

between parents with yellow seeds. The factor, or gene, for the green color must be present but hidden in the parents. For example, a parent with Yy genes would have yellow seeds because the dominant gene (Y) was present. But that parent would also be carrying the recessive gene (y) for green seeds. The green seed trait would be hidden in the parent but could be passed to its offspring.

When the parent plant forms sex cells (sperm or eggs), the parent's gene pairs segregate, or separate. This process is known as the law of segregation. According to the law of segregation, one gene from each pair goes to each sex cell. Half of the sex cells of a hybrid pea plant with the gene pair Yy have a gene for yellow seeds (Y). The other half of the sex cells carry a gene for green seeds (y). As a result of sexual reproduction, a male sex cell (sperm) and a female sex cell (egg) unite to form a fertilized egg. Each fertilized egg contains one gene for seed color from each parent, so the gene pair for seed color is formed again.

Mendel also crossed pea plants that differed from one another by two or more traits. The results of these crosses led to the law of independent assortment. The law of independent assortment states that each gene pair for a trait is inherited independently of the gene pairs for all other traits. For example, when a tall plant with yellow seeds forms sex cells, the genes for stem length separate independently from the genes for seed color.

Figure 1–11 *The diagram illustrates the law of segregation, describing how one of the parent's genes goes to each sex cell. Notice how the baby rhinoceros and the sheepdog puppies resemble their parents. Why do organisms resemble their parents?*

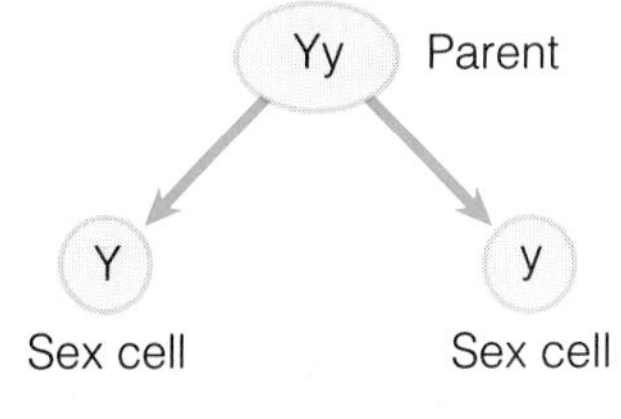

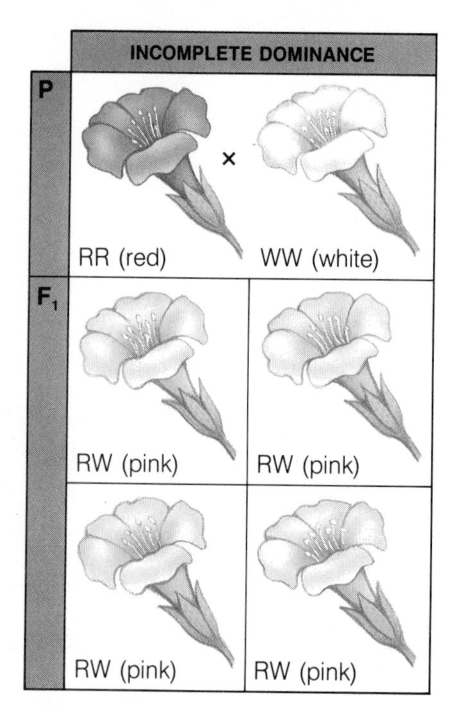

Figure 1–12 *In four-o'clock flowers, neither the red gene nor the white gene for flower color is dominant. When these two genes are present in the same plant, a pink flower results. What kind of inheritance does this show?*

Incomplete Dominance

Mendel's idea that genes are always dominant or always recessive proved to be true in most cases but not all. In 1900, the German botanist Karl Correns made an important discovery. Correns found that in some gene pairs, the genes are neither dominant nor recessive. Instead, these genes show **incomplete dominance.** Incomplete dominance means that neither gene in a gene pair masks the other. As a result, the traits carried by the two genes appear to be blended.

Correns did much of his work with a type of plant called a four-o'clock, shown in Figure 1–12. He discovered that when he crossed purebred red four-o'clock flowers (RR) with purebred white four-o'clock flowers (WW), the result was all pink four-o'clock flowers (RW). Neither the gene for red color nor the gene for white color was dominant. Neither gene in the gene pair was masked. Instead, the two traits for flower color, red and white, seemed to have blended together. This blending resulted in pink flowers. Notice that in the case of incomplete dominance, the symbols for the gene pairs for red flowers, white flowers, and pink flowers are all capital letters. This is because neither the gene for red flowers nor the gene for white flowers is dominant over the other.

Incomplete dominance also occurs in animals. One of the most famous examples of this is seen in the beautiful horses known as palominos. Palominos are pale golden-brown with a white mane and tail. If

Figure 1–13 *Traits—such as the color, markings, tail length, and fur length in these cats—are passed on from one generation of organisms to the next.*

a purebred chestnut-brown horse (BB) is crossed with a purebred creamy-white horse (WW), their offspring will all be palominos (BW). If two palominos are crossed, what colors will their offspring be? Is it possible to have purebred palominos?

Genetic Principles

Through the work of scientists such as Mendel and Correns, certain basic principles of genetics have been established. These basic principles are

- **Traits, or characteristics, are passed on from one generation of organisms to the next generation.**
- **The traits of an organism are controlled by genes.**
- **Organisms inherit genes in pairs, one gene from each parent.**
- **Some genes are dominant, whereas other genes are recessive.**
- **Dominant genes hide recessive genes when both are inherited by an organism.**
- **Some genes are neither dominant nor recessive. These genes show incomplete dominance.**

In 1866, Mendel published a paper describing his experiments and conclusions in a little-known scientific journal. Scientists at that time, however, were not prepared to accept Mendel's results. They did not understand the importance of his work. Mendel's paper remained unread and unappreciated for many years. Finally, in the early 1900s, Mendel's paper was rediscovered. Scientists realized that Mendel had correctly described the basic principles of genetics. Mendel's discoveries were at last recognized as an important scientific breakthrough.

Incomplete Dominance

In this activity, you will use coins to model a cross between two plants with the gene pair RW.

1. Obtain two coins.

2. Cut four equal pieces of masking tape to fit on the coins without overlapping.

3. Place a piece of tape on each side of both coins.

4. Write the letter R on one side of each coin and the letter W on the other side.

5. Toss both coins 100 times and record the gene pair for each toss.

What is the percentage of occurrence for each gene pair?

1–2 Section Review

1. List six basic principles of genetics.
2. What is incomplete dominance?
3. What is a hybrid organism?

Critical Thinking—*Relating Concepts*

4. Can a short-stemmed pea plant ever be a hybrid? Explain why or why not.

Guide for Reading

Focus on this question as you read.

- *How can you predict the results of genetic crosses using probability and Punnett squares?*

1–3 Genetics and Probability

In one of Mendel's experiments, he crossed two plants that were hybrid for yellow seeds (Yy). When he examined the plants that resulted, he discovered that about one seed out of every four was green. By applying the concept of probability to his work, Mendel was able to express his observations mathematically. He could say that the probability of such a cross producing green seeds was 1/4, or 25 percent. Probability is the possibility, or likelihood, that a particular event will take place. **Probability can be used to predict the results of genetic crosses.**

Probability

Activity Bank

Flip Out!, p. 105

Suppose that you are about to toss a coin. What are the chances that the coin will land heads up? If you said a 50-percent chance, you are correct. What are the chances that the coin will land tails up? Again, the answer is 50 percent. Although you may not realize it, you, like Gregor Mendel, used the laws of probability to arrive at your answers. You figured out the chance, or likelihood, that the coin would come up heads (or tails) on one toss.

A probability is usually written as a fraction or as a percentage. For example, the chance that a sex cell will receive a Y gene from a parent with a Yy gene

Figure 1–14 *According to the law of probability, a coin will land heads up 50 percent of the time and tails up 50 percent of the time. What is the probability that the next child born to these parents, who already have four sons, will be a boy? A girl?*

pair is 1/2, or 50 percent. In other words, you would expect one half, or 50 percent, of the sex cells to receive a Y gene.

In probability, the results of one event do not affect the results of the next. Previous events do not affect future outcomes. Each event happens independently. For example, suppose you toss a coin 10 times and it lands heads up each time. What is the probability that it will land heads up on the next toss? Because the coin landed heads up on the previous 10 tosses, you might think that it is also likely to land heads up on the next toss. But this is not the case. The probability of the coin's landing heads up on the next toss is still 1/2, or 50 percent. The results of the first 10 tosses do not influence the result of the eleventh toss.

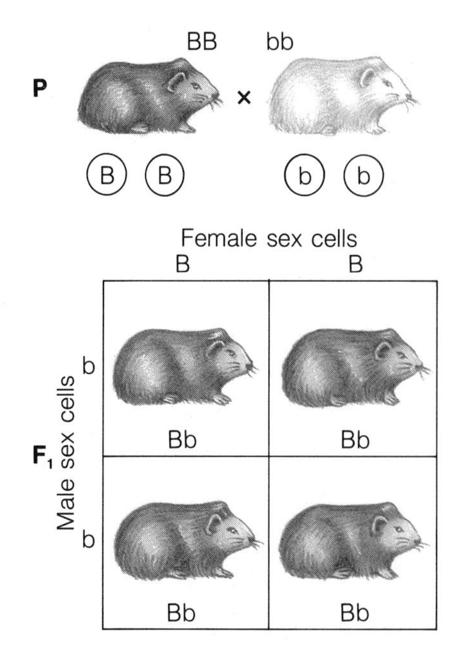

Figure 1–15 *This Punnett square shows a cross between a black guinea pig and a white guinea pig. What is the phenotype of the offspring?*

Punnett Squares

In addition to probability, a special chart called a Punnett square is used to show the possible gene combinations in a cross between two organisms. This chart was developed by Reginald C. Punnett, an English geneticist.

Let's see how a Punnett square works. Look at the Punnett square in Figure 1–15. It shows a cross between two guinea pigs. The two possible genes in the female sex cells are listed across the top of the chart. The two possible genes in the male sex cells are listed along the left side. Remember, when a male sex cell (sperm) and a female sex cell (egg) join, a fertilized egg forms. Each box in the Punnett square represents a possible gene pair in the fertilized egg.

Notice that in the P (parent) generation, both of the female's genes are for black hair (BB). Both of the male's genes are for white hair (bb). All of the offspring in the F_1 generation are hybrid black (Bb). If you were to look at these hybrid black-haired guinea pigs, you would not be able to tell the difference between them and purebred black-haired guinea pigs. Their **phenotypes** (FEE-noh-tighps), or physical appearances, are the same. Phenotype refers to a visible characteristic; in this case, black hair. However, their **genotypes** (JEHN-uh-tighps) are different. A genotype is the actual gene makeup of an organism.

ACTIVITY — CALCULATING

Probability

What is the probability that two plants with the genotype Rr, where R = smooth seeds and r = wrinkled seeds, will produce an offspring that has wrinkled seeds? Express your answer as a fraction and as a percentage.

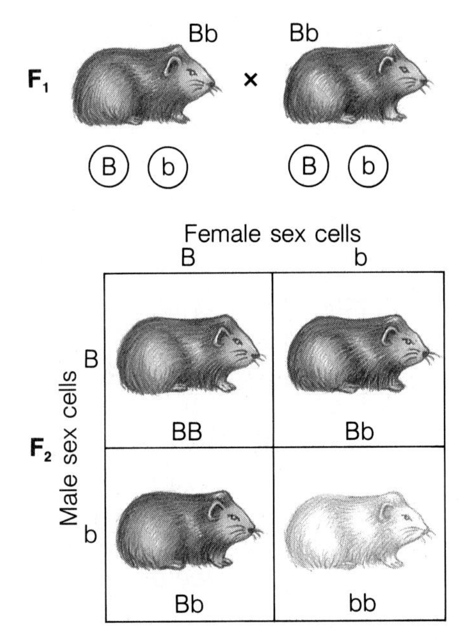

Figure 1–16 *This Punnett square shows a cross between two hybrid black guinea pigs. What phenotypes are present in the offspring?*

A purebred black-haired guinea pig has the genotype BB; a hybrid black-haired guinea pig has the genotype Bb. Both the purebred guinea pig and the hybrid guinea pig have black hair. Figure 1–16 shows a cross between two of the hybrid black-haired guinea pigs from the F_1 generation. The results of the F_1 cross, or the F_2 generation, are 1/4 purebred black (BB), 1/2 hybrid black (Bb), and 1/4 purebred white (bb). What percentage of the offspring have the same genotype as the parents? What percentage have the same phenotype?

1–3 Section Review

1. What is probability?
2. What is the difference between genotype and phenotype? Give an example.

Critical Thinking—*Making Predictions*

3. Use a Punnett square to predict the outcome of a cross between a hybrid black guinea pig and a white guinea pig. What are the possible genotypes of the offspring? What are the possible phenotypes? (*Hint:* What is the gene makeup of the white guinea pig?)

PROBLEM Solving

Using Punnett Squares to Solve Genetics Problems

One-Factor Crosses: Sample Problem

In pea plants, round seeds are dominant over wrinkled seeds. Predict the genotypes and phenotypes of a cross between two hybrid round-seeded pea plants.

SOLUTION:

Step 1 Choose a letter to represent the genes in the cross.

R = round
r = wrinkled

Problem Solving

Use a letter whose capital form does not look too similar to its lowercase form. This will make it easier for you to read your finished Punnett square. Except for that requirement, it is not important which letter you select. In this case, R is the dominant round gene and r is the recessive wrinkled gene.

Step 2 Write the genotypes of the parents.

Rr × Rr

This step is often written as an abbreviation of the cross being studied. The x between the parents' genotypes is read "is crossed with." In this case, Rr is crossed with Rr.

Step 3 Determine the possible genes that each parent can produce.

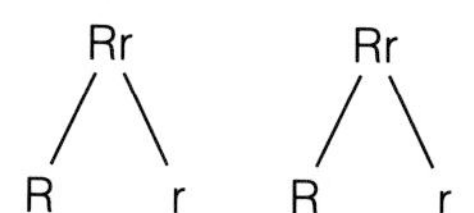

Step 4 Write the possible genes at the top and side of the Punnett square.

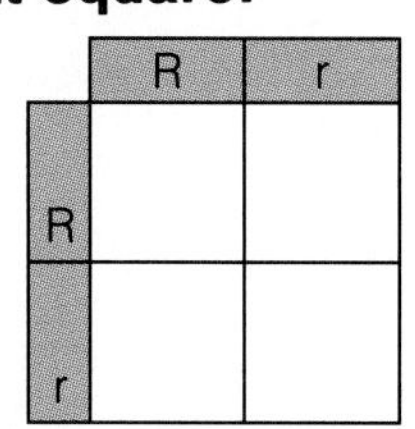

Step 5 Complete the Punnett square by writing the gene combinations in the appropriate boxes.

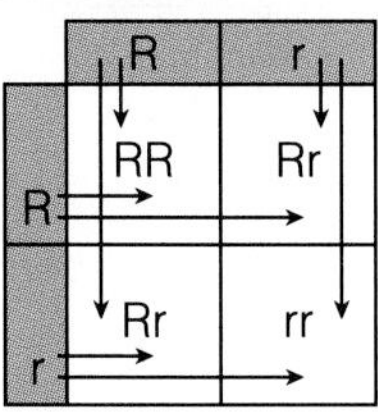

This step represents the process of fertilization. The gene from the top of the box and the gene from the side of the box are combined inside each of the four boxes. If there is a combination of capital letter and lowercase letter in a box, write the capital letter first. The letters inside the box represent the probable genotypes of the offspring resulting from the cross. In this example, 1/4 of the offspring are genotype RR, 1/2 are Rr, and 1/4 are rr.

Step 6 Determine the phenotypes of the offspring.

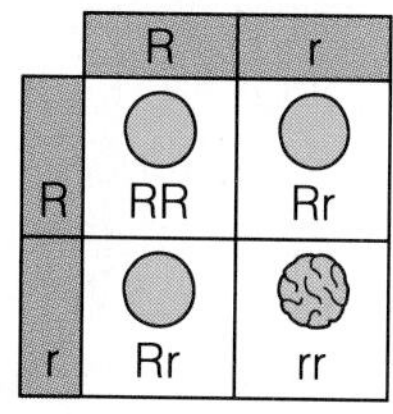

Remember that phenotype refers to the physical appearance of an organism. The principle of dominance makes it possible to determine the phenotype that corresponds to each genotype inside the Punnett square. In this example, 3/4 of the offspring have round seeds and 1/4 of the offspring have wrinkled seeds.

One-Factor Crosses: Practice Problems

1. What are the possible genotypes and phenotypes produced when a hybrid round-seeded pea plant (Rr) is crossed with a purebred wrinkled-seeded plant (rr)?

2. What are the possible genotypes and phenotypes produced when a purebred round-seeded pea plant (RR) is crossed with a hybrid round-seeded plant (Rr)?

Laboratory Investigation

Dominant and Recessive Traits

Problem

What are the phenotypes of some dominant and recessive human traits?

Materials *(per group)*

paper
pencil

Procedure

1. Copy the data table shown here on a sheet of paper.
2. Count the number of students in your class who have each of the traits listed in the table. Tongue rolling is the ability to roll the tongue into a U-shape. Free ear lobes are those that hang below the point of attachment to the head. Attached ear lobes are attached directly to the side of the head. A widow's peak is a distinct point in the hairline in the center of the forehead. Record your results in the data table.
3. Determine the percentage of students who demonstrate each trait as follows: Divide the number of students who have the trait by the total number of students in your class and multiply by 100. Record the percentages in the data table.

Observations

1. Which trait is most common in your class? Which is least common?
2. Do any students have traits that are intermediate between the dominant and recessive traits? If so, how many? Describe the intermediate traits.

Analysis and Conclusions

1. Do dominant traits occur more often than recessive traits? Explain.
2. Predict how your observations might be different if you were to observe four other classes of students.
3. **On Your Own** State a hypothesis to explain why some students have traits that are intermediate.

DATA TABLE

Traits		Number of Students Demonstrating Dominant Trait	Number of Students Demonstrating Recessive Trait	Percentage Demonstrating Dominant Trait	Percentage Demonstrating Recessive Trait
Dominant	*Recessive*				
Tongue roller	Nonroller				
Free ear lobes	Attached ear lobes				
Dark hair	Light hair				
Widow's peak	Straight hairline				
Nonred hair	Red hair				
Total					

Study Guide

Summarizing Key Concepts

1–1 History of Genetics

▲ Genetics is the study of heredity, or the passing on of traits from an organism to its offspring.

▲ For each trait, every organism has a pair of factors, or units of heredity, called genes.

▲ The stronger of two genes for a trait is called the dominant; the weaker is called the recessive.

1–2 Principles of Genetics

▲ A purebred organism has genes that are alike for a particular trait (TT or tt).

▲ A hybrid has genes that are different for a trait (Tt).

▲ According to the law of segregation, one gene from each gene pair goes to each sex cell.

▲ The law of independent assortment states that each gene pair is inherited independently of the gene pairs for all other traits.

▲ In some gene pairs, the genes show incomplete dominance; that is, neither gene hides the other.

1–3 Genetics and Probability

▲ Probability can be used to predict the results of genetic crosses.

▲ Probability is the chance, or likelihood, that an event will happen.

▲ In probability, the results of one event do not affect the results of the next event.

▲ Punnett squares show the possible gene combinations resulting from a cross between two organisms.

▲ A phenotype describes a visible characteristic, whereas a genotype is the actual gene makeup.

Reviewing Key Terms

Define each term in a complete sentence.

1–1 History of Genetics

trait
genetics
gene
dominant
recessive

1–2 Principles of Genetics

hybrid
incomplete dominance

1–3 Genetics and Probability

phenotype
genotype

Chapter Review

Content Review

Multiple Choice

Choose the letter of the answer that best completes each statement.

1. The male reproductive structures of pea plants are called
a. pistils. c. petals.
b. stamens. d. pollen.

2. Who is called the Father of Genetics?
a. Watson c. Mendel
b. Correns d. Punnett

3. The symbol for a dominant gene is written as
a. a capital letter.
b. a lowercase letter.
c. a capital letter and a lowercase letter.
d. two lowercase letters.

4. When Mendel crossed two short pea plants, the offspring were
a. all short. c. 1/2 short and 1/2 tall.
b. all tall. d. 3/4 tall and 1/4 short.

5. Mendel studied all of the following traits of pea plants except
a. stem length. c. flower color.
b. seed color. d. pod color.

6. Gene pairs for a trait separate according to the law of
a. independent assortment.
b. incomplete dominance.
c. hybridization.
d. segregation.

7. Which gene pair would a hybrid tall pea plant have?
a. TT c. Tt
b. tt d. none of these

8. The probability that a pea plant will receive a T gene from a Tt parent is
a. 1/4. c. 50 percent.
b. 3/4. d. 100 percent.

True or False

If the statement is true, write "true." If it is false, change the underlined word or words to make the statement true.

1. One reason Mendel studied pea plants is that they grow and reproduce <u>slowly</u>.

2. The <u>pistil</u> of a pea plant flower produces pollen.

3. The process by which a plant pollinates itself is called <u>cross-pollination</u>.

4. Mendel called plants that always produce offspring with the same traits as the parents <u>true-breeding</u> plants.

5. Two of Mendel's laws of genetics are the law of <u>probability</u> and the law of independent assortment.

6. Probability is usually expressed as a fraction or as a <u>percentage</u>.

7. Scientists who study heredity are called <u>plant breeders</u>.

Concept Mapping

Complete the following concept map for Section 1–1. Refer to pages E6–E7 to construct a concept map for the entire chapter.

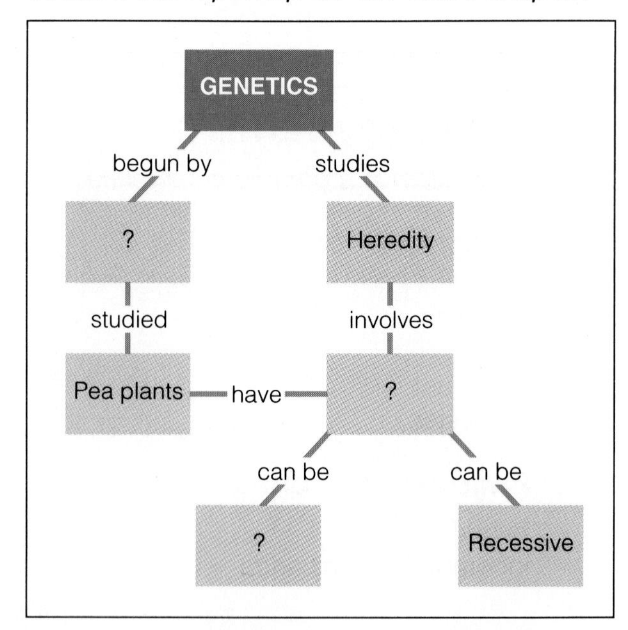

Concept Mastery

Discuss each of the following in a brief paragraph.

1. Explain why Gregor Mendel chose pea plants for his experiments.
2. Describe what happened when Mendel crossed two tall pea plants. How would Mendel's conclusions have been different if he had studied only one generation of pea plants?
3. Why was the importance of Mendel's work not recognized until the early 1900s?
4. What is the difference between cross-pollination and self-pollination?
5. In your own words, explain the law of segregation and the law of independent assortment, using specific examples.
6. Why does an offspring have a 50 percent chance of receiving a B gene from a parent with a Bb gene pair for hair color?

Critical Thinking and Problem Solving

Use the skills you have developed in this chapter to answer each of the following.

1. **Making predictions** A family has four daughters. What is the probability that a fifth child will be a girl? Does the fact that there are already four daughters in the family increase the probability of having another girl? Explain.
2. **Making diagrams** Complete the Punnett square to show the possible genotypes of the F_1 generation when a hybrid black (Bb) rabbit is crossed with a purebred brown (bb) rabbit.

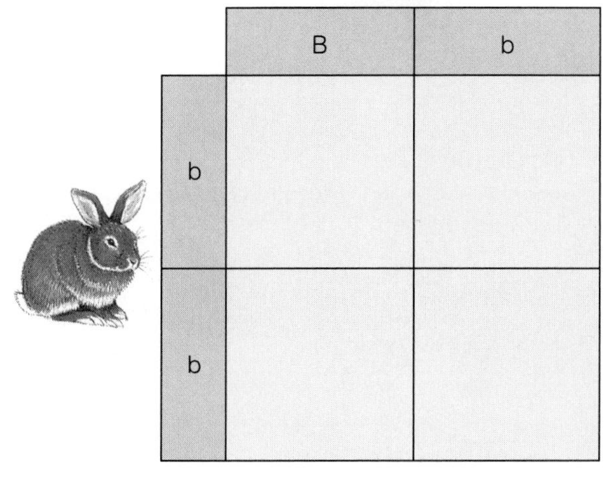

3. **Applying concepts** In pea plants, short stems (t) are recessive and tall stems (T) are dominant. Two hybrid pea plants (Tt × Tt) are crossed. One hundred seeds from the two plants are collected and planted. How many plants in the next generation would you expect to have tall stems?
4. **Designing an experiment** Pink four-o'clock flowers (RW) can be produced by crossing a plant with red flowers (RR) and a plant with white flowers (WW). Design an experiment to produce a plant with red flowers from a plant with pink flowers. Draw a Punnett square to illustrate the results of your experiment.
5. **Making calculations** A short-tailed cat (LS) is crossed with a long-tailed cat (LL). If six kittens are born, how many might have short tails? How many might have long tails? Draw a Punnett square to illustrate this cross. Could any of the kittens have no tails? Explain.
6. **Using the writing process** Imagine that you are a student in the 1860s visiting Gregor Mendel in his garden. Write a letter to a friend describing Mendel's experiments with pea plants.

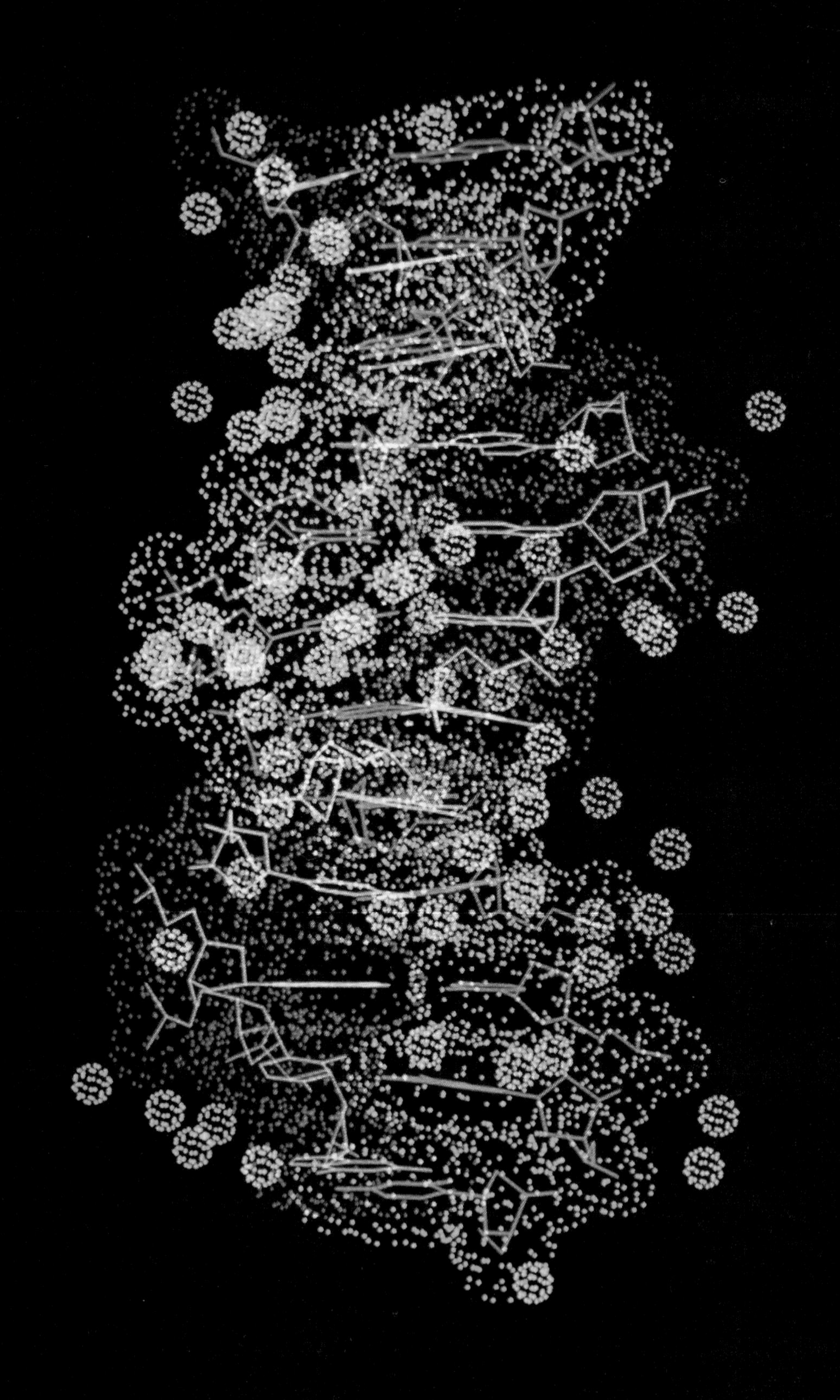

How Chromosomes Work

Guide for Reading

After you read the following sections, you will be able to

2–1 The Chromosome Theory

- Explain what is meant by the chromosome theory of heredity.
- Describe how the sex of an organism is determined.

2–2 Mutations

- Define mutations.

2–3 The DNA Molecule

- Describe the structure of the DNA molecule.
- Explain how a DNA molecule replicates.

2–4 How Chromosomes Produce Proteins

- Describe the process of protein synthesis.

Like a strange galaxy in outer space, the atoms in a computer model of a DNA molecule sparkle against a black background. In many ways, the DNA molecule is similar to a galaxy. Both are made up of many smaller parts. One helps explain the nature of the universe. The other helps explain the nature of living things.

DNA is sometimes called the "code of life." Hidden in the structure of the DNA molecule is the genetic code that shapes every living thing. Because no two living things contain exactly the same code, all living things are different. Variations in the genetic code result in the wonderful diversity of life.

Scientists have unraveled the way in which the chemical instructions in DNA are passed on from one generation to the next. In this chapter, you will learn how the knowledge locked inside the DNA molecule was decoded. Using this knowledge, scientists are now able to change the instructions in some DNA molecules. In doing so, they have produced changes in various forms of life—changes that have never existed before. These forms of life have already helped humans in ways we never dreamed of.

Journal *Activity*

You and Your World Genetic information is transmitted by means of "code words" in your genes. Try making up your own secret code. In your journal, write a short message in code. See if a classmate can translate your coded message.

This computer graphics image of a DNA molecule shows its characteristic spiral structure.

Guide for Reading

Focus on these questions as you read.

- *What is meant by the chromosome theory of heredity?*
- *What is the main function of chromosomes?*

2–1 The Chromosome Theory

As you learned in Chapter 1, the work of Gregor Mendel provided many early solutions to the riddle of genetics. But Mendel did not have all the answers. For example, he did not know where the hereditary factors, or genes, are located in the cell. By the time Mendel's work was rediscovered, however, scientists had more clues to work with and better tools to help them in their research. The first clue came in 1882 when the German biologist Walther Flemming discovered **chromosomes.** Chromosomes are rod-shaped structures that are found in the nucleus of every cell in an organism. The next clue was provided by Walter Sutton, an American graduate student who in 1902 was doing research on chromosomes. While observing grasshopper chromosomes, Sutton discovered where the genes in a cell are located.

Grasshoppers have 24 chromosomes, arranged in 12 pairs. This means that every body cell in a grasshopper contains 24 chromosomes. Sutton observed that each of a grasshopper's sex cells (sperm or egg) contained 12 chromosomes, or half the number in body cells. Sutton also observed what happened when a male sex cell (sperm with 12 chromosomes) and a female sex cell (egg with 12 chromosomes)

Figure 2–1 *The photograph on the right shows the chromosomes of a mouse magnified 4000 times. What is the main function of chromosomes?*

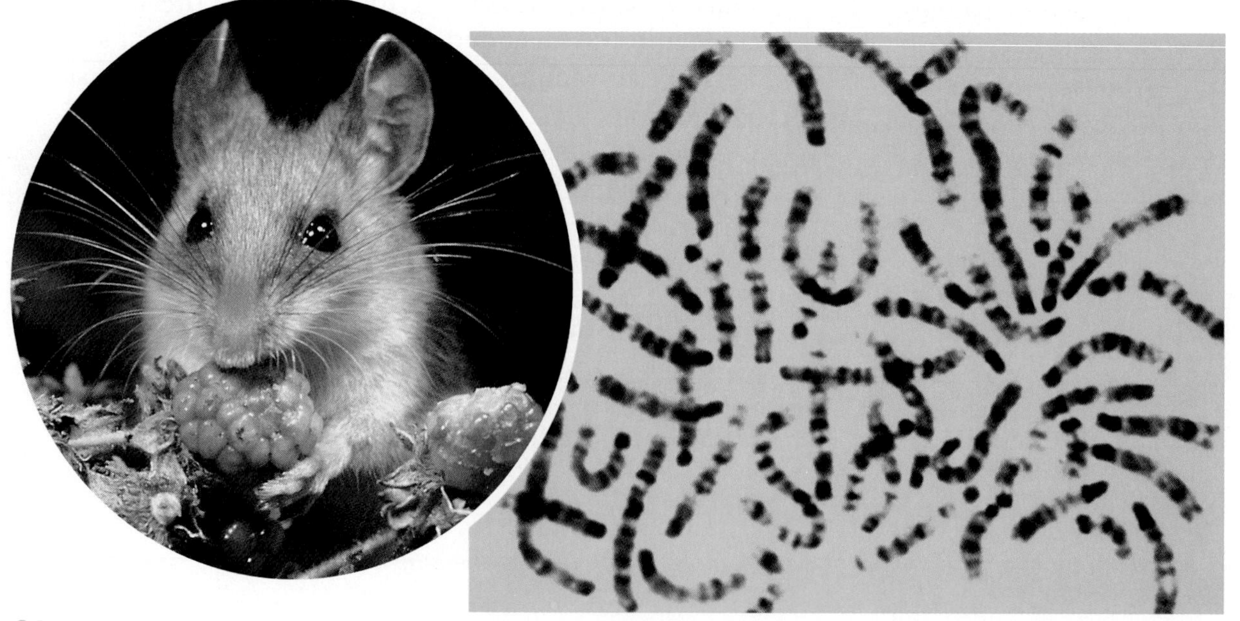

joined. The fertilized egg that was formed had 24 chromosomes, the original number, arranged in 12 pairs. In other words, the grasshopper offspring had exactly the same number of chromosomes as each of its parents.

From his work, Sutton concluded that chromosomes carried Mendel's hereditary factors, or genes, from one generation to the next. In other words, genes are located on chromosomes. Sutton's idea that genes are found on chromosomes came to be known as the chromosome theory of heredity. **According to the chromosome theory, genes are carried from parents to their offspring on chromosomes.** How amazing it now seems that Mendel was able to do all of his work without even knowing about chromosomes!

Scientific Serendipity

In your study of science, you will find that many important discoveries were made by accident. Serendipity means being able to find something valuable without looking for it. All of the scientists listed below made important discoveries in this way. Use library references to find out what their discoveries were and how chance or accident helped in those discoveries. Write a report of your findings.

Alexander Fleming
Henri Becquerel
Friedrich Kekule
Arno Penzias and Robert Wilson

Chromosomes and Genes

Today, scientists know that chromosomes play an essential role in heredity. Chromosomes control all the traits of an organism. How do they perform this complex task? The main function of genes on chromosomes is to control the production of substances called proteins. All organisms are made up primarily of proteins. Proteins determine the size, shape, and other physical characteristics of an organism. In other words, proteins determine the traits of an organism. The kind and number of proteins in an organism determine the traits of that organism. So by controlling the kind and number of proteins produced in an organism, chromosomes are able to determine the traits of that organism.

Chromosomes are found in pairs within the nucleus of a cell. Generally, for any particular trait, the gene contributed by one parent is on one of the paired chromosomes. The other gene for that trait, contributed by the other parent, is on the second chromosome of the pair. Each gene's major role is to control the production of a specific protein. So a chromosome, which contains many genes, actually controls the production of a wide variety of proteins. In the last section of this chapter, you will learn more about how chromosomes control the production of proteins.

Figure 2–2 *Like most of the rest of the body, human hair is made of protein. This photograph of human hair was taken with a special microscope called a scanning electron microscope.*

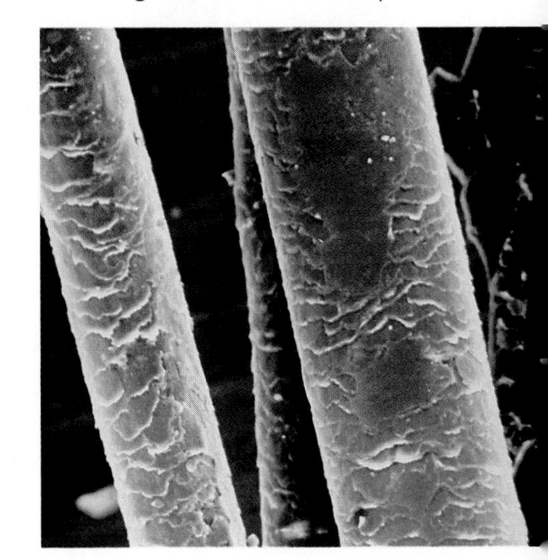

Figure 2–3 *According to the chromosome theory, genes are located on chromosomes, which occur in pairs. Cobras have 38 chromosomes. How many pairs of chromosomes does a cobra have?*

Figure 2–4 *The law of segregation states that the two genes for a trait are separated during the formation of sex cells. During meiosis, chromosomes double and then separate. Different gene pairs on different chromosomes separate independently.*

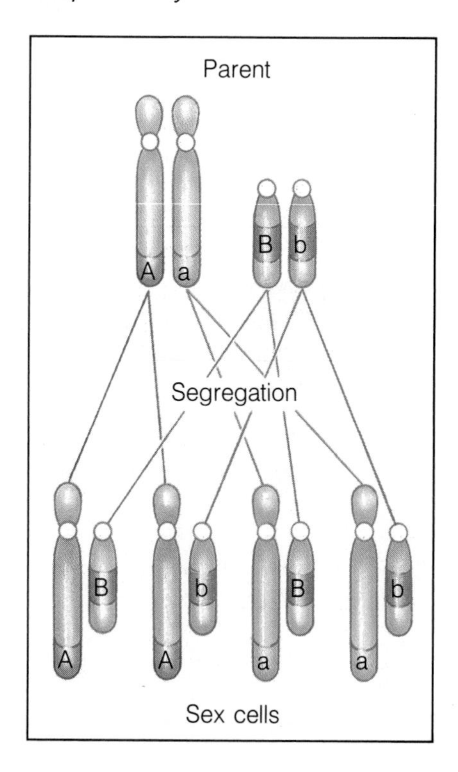

Meiosis

Exactly how are chromosomes passed on from parents to offspring? After all, if each parent contributed all of its chromosomes to an offspring, then the offspring would have twice as many chromosomes as its parents—twice the normal number of chromosomes. This does not happen because of a process called **meiosis** (migh-OH-sihs).

The process of meiosis produces the sex cells, the sperm or egg cells. Remember that according to the law of segregation described in Chapter 1, each of an organism's two genes for a particular trait are separated, or segregated, during the formation of sex cells. This is precisely what happens during meiosis. As a result of meiosis, the number of chromosomes (and genes they carry) in each sex cell is half the normal number of chromosomes found in the parent. When sex cells combine to form the offspring, each sex cell contributes half the normal number of chromosomes. Thus, the offspring gets the normal number of chromosomes—half from each parent.

You can see in Figure 2–5 how meiosis works. In this example, each parent cell has four chromosomes. The chromosomes are arranged in two pairs.

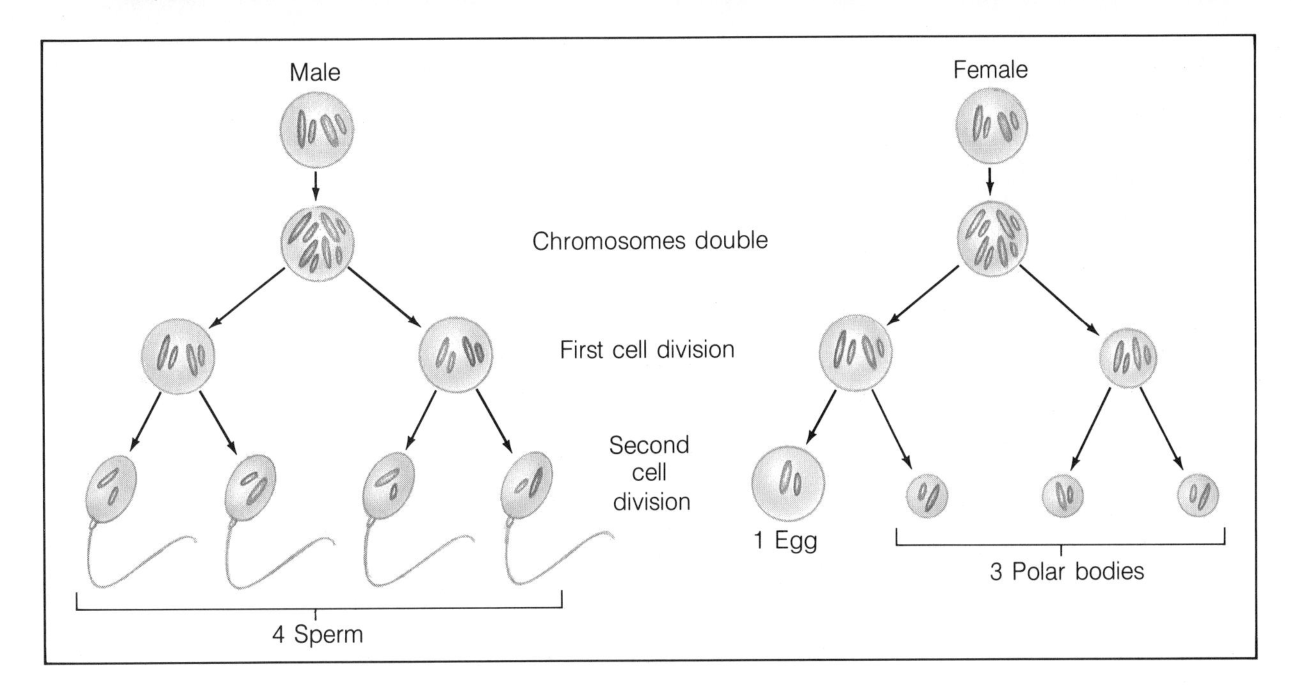

Refer to the diagram as you read the following description of meiosis.

The first thing that happens during meiosis is that the chromosomes in the cell double, producing eight chromosomes. Then the cell divides. During this cell division, the chromosome pairs separate and are equally distributed. So each of the two cells formed by this cell division has four chromosomes, the original number. Next, these two cells divide. Each of the resulting four cells now has two chromosomes. That is, each cell in the last group of cells produced by meiosis has half the number of chromosomes as the original parent cell.

The table in Figure 2–6 illustrates the relationship between Sutton's discoveries about chromosomes and Mendel's conclusions about hereditary factors, or genes. Can you see from this table what led Sutton to conclude that genes must be carried on chromosomes?

Figure 2–5 *During the process of meiosis, a male or a female cell undergoes two divisions, resulting in sex cells (sperm and egg) that have half the normal number of chromosomes. The polar bodies disintegrate, leaving only one egg cell.*

A Model of Meiosis, p.106

Figure 2–6 *This table shows how a comparison of the work of Mendel and Sutton supports the chromosome theory of heredity. What is the chromosome theory?*

COMPARISON OF MENDEL'S AND SUTTON'S DISCOVERIES

Mendel's Factors (genes)	Sutton's Chromosomes
Genes occur in pairs	Chromosomes occur in pairs
Genes in a pair separate	Chromosomes separate
Sex cells contain one gene from each gene pair	Sex cells contain half the normal number of chromosomes

Sex Chromosomes

In 1907, the American zoologist Thomas Hunt Morgan began his own studies in genetics. He experimented with tiny insects called fruit flies. You may have seen fruit flies hovering over the fruits and vegetables in a grocery store or supermarket. Morgan chose to study fruit flies for three reasons. First, fruit flies are easy to raise. Second, they produce new generations of offspring very quickly. Third, their body cells have only four pairs of chromosomes (eight chromosomes), making them easy to study.

Morgan quickly discovered something strange about the fruit flies' four pairs of chromosomes. In female fruit flies, the chromosomes of each pair were the same shape. In males, however, the chromosomes of one pair were not the same shape. One chromosome of the pair was shaped like a rod, and the other chromosome of the pair was shaped like a hook. Morgan called the rod-shaped chromosome the X chromosome and the hook-shaped chromosome the Y chromosome.

After performing a number of experiments and analyzing his results, Morgan discovered that the X and Y chromosomes determine the sex of an organism. For this reason, the X and Y chromosomes are called **sex chromosomes.** In general, an organism (such as a fruit fly or a human) that has two X chromosomes (XX) is a female. An organism that has one X chromosome and one Y chromosome (XY) is a male. There are some exceptions to this general rule. Female birds, for example, have an X chromosome and a Y chromosome instead of two X chromosomes. Based on this information, can you predict what sex chromosomes are found in male birds?

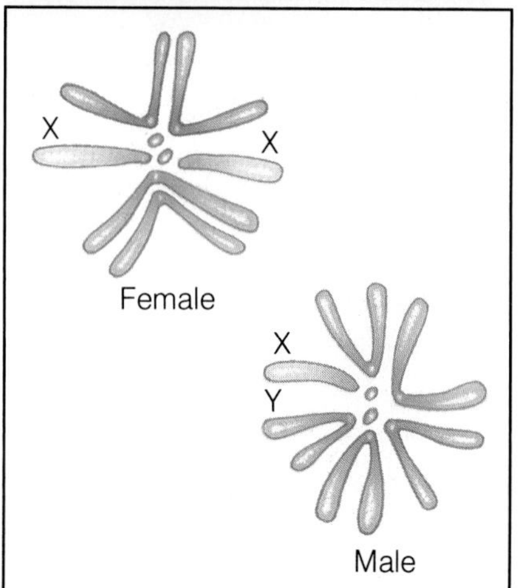

Figure 2–7 *A fruit fly, shown here magnified ten times, has four pairs of chromosomes. A female fruit fly has two X chromosomes. A male fruit fly has one X and one Y chromosome. What are the X and Y chromosomes called?*

Activity Bank

Stalking the Wild Fruit Fly, p.108

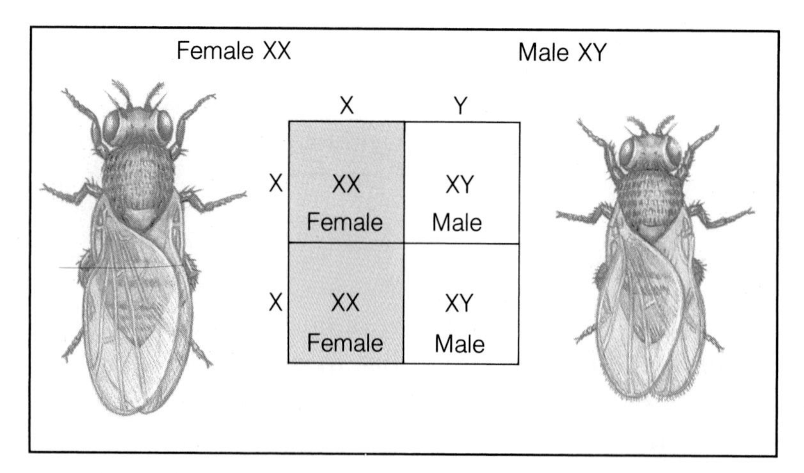

Figure 2–8 *You can see from the drawings of the male and female fruit flies that they have some physical differences. The Punnett square shows the probable sex of their offspring.*

2–1 Section Review

1. State the chromosome theory of heredity. What is the main role of chromosomes in heredity?
2. What is meiosis?
3. Why are the X and Y chromosomes called the sex chromosomes?

Critical Thinking—*Sequencing Events*

4. In your own words, describe the sequence of steps in the process of meiosis. How does this process explain Mendel's law of segregation?

2–2 Mutations

Guide for Reading

Focus on this question as you read.

▶ *What are mutations?*

In 1886, the Dutch botanist Hugo De Vries (duh-VREEZ) made an accidental discovery. The results of his discovery would take the science of genetics beyond the ground-breaking work of Gregor Mendel. De Vries was out walking one day when he came across a group of flowers called American evening primroses. As with Mendel's pea plants, some primroses appeared very different from others. De Vries wondered why this was so. He bred the primroses and got results similar to the results of Mendel's experiments with pea plants. But he also found that every once in a while, new variations appeared among the primroses—variations that could not be explained by the laws of genetics at that time.

Genetic Mistakes

De Vries called the sudden changes he observed in the characteristics of primroses **mutations.** Mutations are genetic mistakes that can affect the way in which traits are inherited. The word mutation comes from a Latin word that means change. **A mutation is a change in a gene or chromosome.** If a gene or chromosome mutation occurs in a body cell such as a skin cell, the mutation affects only the organism that carries it. But if a mutation takes place in a sex cell, that mutation can be passed on to

Figure 2–9 *In chromosome mutations, part of a chromosome may be lost (top), turned around (center), or become attached to a different chromosome (bottom). What may happen if one of these mutations takes place in a sex cell?*

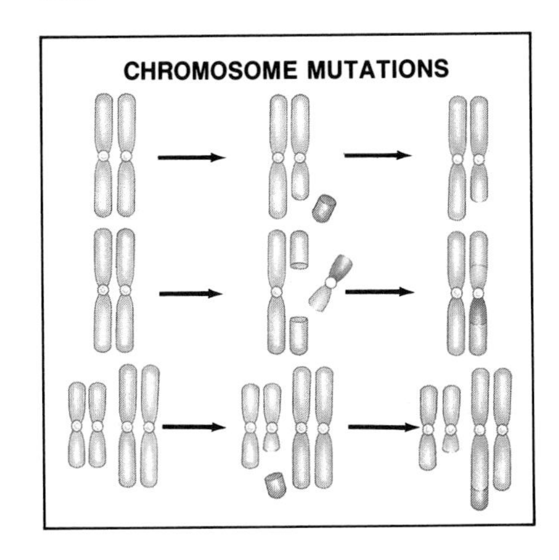

Figure 2–10 *The striking appearance of this buck, or male deer, is the result of a mutation in coat color. The white baby alligator (shown with a normal baby alligator) is also the result of a mutation. What is a mutation?*

an offspring. The mutation may then cause a change in the characteristics of the next generation.

Harmful Mutations

Many mutations are harmful; that is, they reduce an organism's chances for survival or reproduction. For example, one mutation in a gene causes a serious human blood disease called sickle cell anemia. Sickle cell anemia results in red blood cells that are shaped like a crescent moon (or like a sickle, which is a farm tool used to cut grain). Red blood cells normally carry oxygen to the body cells. People who have two genes for sickle cell anemia have difficulty obtaining enough oxygen because the sickle-shaped red blood cells cannot carry enough of this vital gas to all the cells in the body. Sickle cell anemia also causes severe pain, because the sickle-shaped cells may clump together and clog tiny blood vessels. If left untreated, sickle cell anemia may cause death.

Helpful Mutations

Not all mutations are harmful. Some mutations are helpful and improve an organism's chances for survival. Other mutations are considered "helpful" because they cause desirable traits in organisms that are useful to humans. For example, when mutations occur in crop plants, the crops may become more useful to people. A gene mutation in potatoes produced a new variety,

Figure 2–11 *A mutation has left this frog with six legs. These sweet and juicy navel oranges are also the result of a mutation.*

called the Katahdin potato. This potato is resistant to diseases that attack other varieties. The new potato also looks and tastes better than other types of potatoes. Seedless navel oranges are also the result of a mutation. These oranges are sweeter and juicier than ordinary oranges with seeds.

It may seem to you that mutations produce only helpful or harmful traits. However, this is not so. Many mutations are neutral and do not produce any obvious changes in an organism. Still other mutations are lethal, or deadly, and result in the immediate death of an organism.

Mutagens

Mutations may occur spontaneously (that is, on their own) or they may be caused by some factor in the environment. Factors that cause mutations, such as radiation and certain chemicals, are **mutagens.** Mutagens can be harmful to living things. For example, ultraviolet radiation from the sun damages the genes in skin cells and may cause skin cancer. But mutagens can also be helpful. In fact, scientists have used radiation to sterilize insect pests to prevent them from reproducing. This method was tried in California to control an invasion of the Mediterranean fruit fly, or medfly, which was damaging valuable crop plants.

Mutagens can also be used to speed up the rate of mutations. This technique is often used with

ACTIVITY THINKING

The California Medfly Controversy

The Mediterranean fruit fly, or medfly, is a serious threat to crops in California. State agricultural officials have tried many ways to solve the medfly problem. One method is to spray pesticides that kill the fruit flies. Many citizens' organizations have opposed widespread spraying in their communities. Using newspapers or other reference materials, find out more about this controversy. Form two teams in your class to debate this issue.

CAREERS

Plant Breeder

A **plant breeder** is a scientist who tries to improve plants through genetic methods. Plant breeders may try to improve forests by changing the hereditary information passed on from one generation of trees to the next. Or they may try to develop new crop plants with more desirable traits.

Plant breeders are employed by colleges and universities, government agencies, and the lumber industry, as well as by seed, food, and drug companies. If you are interested in plants, or enjoy growing plants, you may want to learn more about a career as a plant breeder. For more information, write to the U.S. Department of Agriculture, Science and Education, 6505 Belcrest Road, Hyattsville, MD 20782.

Figure 2–12 *Some damaging oil spills can be cleaned up through the use of mutant oil-eating bacteria (inset).*

bacteria. Because bacteria reproduce so rapidly, the use of mutagens increases the chances of producing helpful mutations in the bacteria. Mutant bacteria that can digest oil and are thus useful in cleaning up some oil spills have been produced in this way.

2–2 Section Review

1. What is a mutation? Why are some mutations harmful? Why are some helpful? Give an example of each type of mutation.
2. What is a mutagen? What are some ways in which mutagens are used?
3. Why are mutations that do not occur in sex cells not passed on to future generations?

Connection—*Environmental Science*

4. How can a particular mutation be helpful in preventing environmental pollution?

CONNECTIONS

Making Mutant Mosquitoes

Ever since Thomas Hunt Morgan's experiments with fruit flies, geneticists have used these fast-breeding insects in their research. Today, some researchers are looking at another insect pest as a subject of genetic experiments.

Mosquitoes are responsible for the spread of many diseases, including malaria, encephalitis, and yellow fever. In the past, these diseases were usually attacked by using pesticides to wipe out populations of the disease-carrying mosquitoes or by draining the wetlands where the mosquitoes breed. But both of these approaches resulted in environmental problems. Then in the 1970s, scientists tried to alter the mosquito populations instead of completely eliminating them. They used large doses of radiation to sterilize male malaria-carrying mosquitoes so they would be unable to produce offspring. The scientists hoped that when the laboratory-bred sterilized males were released into the wild, they would mate with wild female mosquitoes. After mating with the sterilized males, the female mosquitoes would not produce offspring, and the mosquito population would decrease.

Unfortunately, these sterilization efforts were only partially successful. One reason was that the doses of radiation used were too large. Instead of just making the mosquitoes sterile, the radiation almost killed them and prevented them from mating at all. Another reason was that when the laboratory-bred male mosquitoes were released into the wild, they were unable to attract wild female mosquitoes!

Scientists have now begun to look for ways to alter only the disease-causing genes, leaving the mosquitoes' other traits untouched. Several laboratories in the United States are trying to map all the genes on the three chromosomes of *Aedes aegypti,* the mosquito which carries the virus that causes yellow fever. They have succeeded in locating several genes that are responsible for the mosquitoes' ability to transmit the disease. The results of these experiments may one day allow scientists to alter the inherited traits of specific mosquito populations and make the mosquitoes incapable of transmitting human diseases.

2–3 The DNA Molecule

Guide for Reading

Focus on this question as you read.

- *What is the role of DNA in heredity?*

"We wish to suggest a structure for the salt of deoxyribose nucleic acid." So began a letter written to a scientific journal by two scientists in 1953. What followed in this letter was a description of the spiral-shaped structure of a molecule that would help unlock the deepest secrets of genetics. This molecule is called **DNA.** DNA stands for **deoxyribonucleic** (dee-AHKS-ih-righ-boh-noo-KLEE-ihk) **acid.** The two scientists who wrote the letter were the American biologist James Watson and the British physicist Francis Crick. In 1962, Watson and Crick (along with another British scientist, Maurice Wilkins) shared the Nobel prize for physiology or medicine for their work on the structure of DNA.

Discovering the Discoverers

James Watson's book *The Double Helix* provides an inside look at the process of scientific discovery.

The Search for the Genetic Code

The DNA molecule is the basic substance of heredity. **DNA stores and passes on genetic information from one generation to the next.** Many scientists believe that the discovery of the structure of DNA was the most important biological breakthrough of the twentieth century.

Several scientists were involved in research on the structure of the DNA molecule. The British scientist Rosalind Franklin, working with Maurice Wilkins, managed to gather large amounts of DNA fibers. When she aimed X-rays at the DNA, she obtained patterns such as the one shown in Figure 2–14. The pattern in these X-ray photographs provided Watson and Crick with important clues about how the DNA molecule was put together. Using models, they were able to come up with a structure for DNA that matched the pattern in Rosalind Franklin's X-ray photographs.

Figure 2–13 *James Watson (left) and Francis Crick (right) are shown in front of their model of a DNA molecule. What do the letters DNA stand for?*

Finding the structure of DNA allowed scientists to crack the genetic code. Watson and Crick's discovery showed that chromosomes are made up of long strands of DNA molecules. It is the DNA molecules in chromosomes that make up the genes. So DNA actually controls the production of the proteins that determine all of the traits passed on from parents to their offspring. You may be wondering

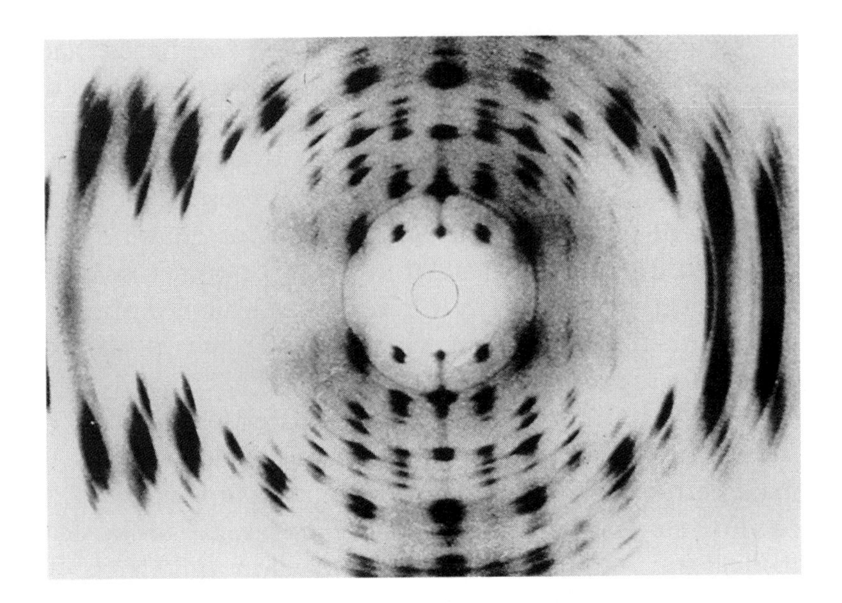

Figure 2–14 *Images of DNA similar to the one shown in this X-ray diffraction photograph helped Watson and Crick discover the structure of the DNA molecule.*

why Rosalind Franklin, whose research played an important part in unlocking the structure of DNA, did not share the 1962 Nobel prize with Watson, Crick, and Wilkins. The reason is that Franklin died in 1958, and Nobel prizes are given only to living scientists.

The Structure of DNA

Figure 2–15 shows the structure of the DNA molecule. A DNA molecule looks like a twisted ladder, or spiral staircase. The sides of the ladder are made up of sugar molecules and phosphate groups (containing the elements hydrogen, phosphorus, and oxygen). The steps, or rungs, of the ladder are formed by pairs of substances called nitrogen bases. Nitrogen bases are molecules that contain the element nitrogen, as well as other elements. There are four different nitrogen bases in DNA. They are adenine (AD-uh-neen), guanine (GWAH-neen), cytosine (SIGHT-oh-seen), and thymine (THIGH-meen). The capital letters A, G, C, and T are used to represent the four different bases.

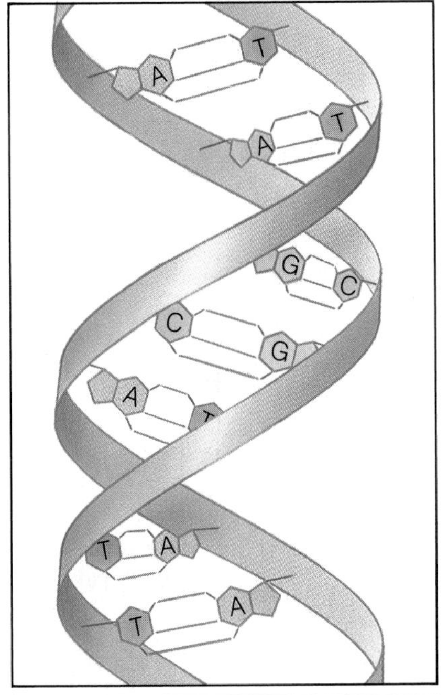

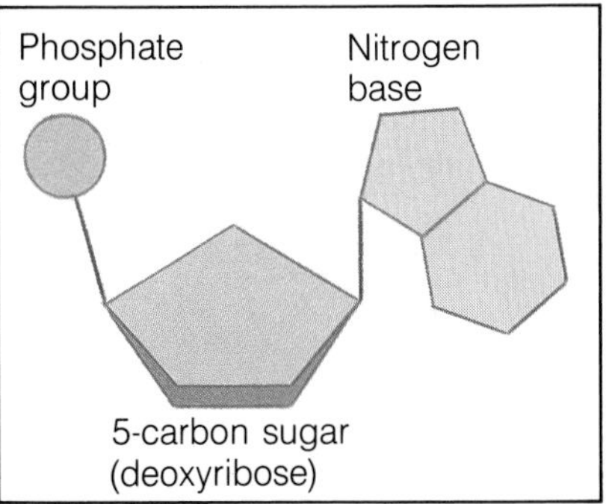

Figure 2–15 *This illustration shows the ladderlike structure of a DNA molecule (top). The DNA molecule is made up of smaller units consisting of a sugar, a phosphate group, and a nitrogen base (bottom). Which of these substances make up the sides of the DNA ladder?*

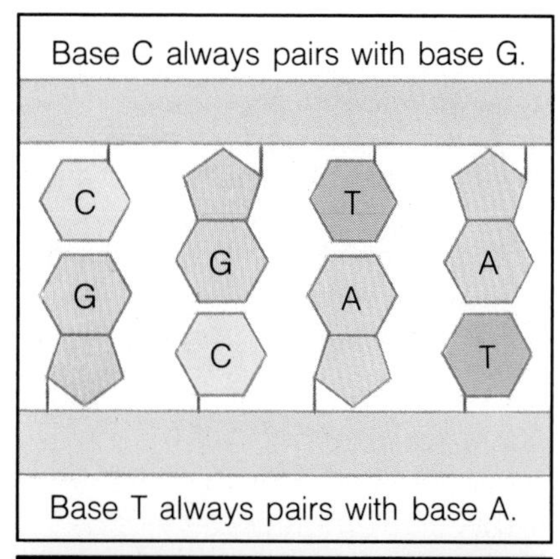

Figure 2–16 *In a DNA molecule, the nitrogen bases that make up the rungs of the ladder always combine in a specific way (top). Which nitrogen base always combines with thymine? The scientist is holding a flask that contains the nitrogen base cytosine. The cytosine is synthetic and was produced in the laboratory.*

As you can see in Figure 2–16, the two bases that make up each rung of the DNA ladder are combined in very specific ways. The American biochemist Erwin Chargaff had found that in any sample of DNA, the amount of adenine is always equal to the amount of thymine. The same is true for guanine and cytosine. Using these clues, Watson and Crick reasoned that in the DNA ladder, adenine (A) always pairs with thymine (T) and guanine (G) always pairs with cytosine (C).

A DNA ladder may contain hundreds or even thousands of rungs. So the DNA molecule that makes up a single chromosome may have hundreds or thousands of pairs of nitrogen bases. In addition to discovering DNA's structure, Watson and Crick reasoned that the order of the nitrogen bases on a DNA molecule determines the particular genes on a chromosome. That is why DNA is said to carry the genetic code. The genetic code is actually the order of nitrogen bases on the DNA molecule—for example, ACGGTTCAAG.

Because a DNA molecule can have many hundreds of bases arranged in any order, the number of different genes is almost limitless. Different genes produce different proteins. That is why living things on Earth can display such a wide variety of traits. Changing the order of only one pair of nitrogen bases in a DNA molecule can result in a new gene that determines a completely different trait. The sequences ATTCGG and TATCGG, for example, differ in the order of only two letters. Yet this small difference might be enough to change the genetic code and thereby produce two totally different proteins, resulting in different traits. In fact, most mutations are actually just a small change in the order of bases in a particular gene.

Cell Reproduction

As an organism grows and develops, the number of body cells must increase. In order for the total number of cells to increase and for the organism to grow, each cell must reproduce. A cell reproduces by dividing into two new cells. Each new cell, called a daughter cell, is identical to the parent cell. As a result, each body cell in an organism contains all the genetic information that determines the organism's traits.

For a parent cell to produce two identical daughter cells, the exact contents of its nucleus must be transferred into the nucleus of each new cell. In other words, the genetic code in the parent cell must be passed on to each daughter cell. Before a body cell can divide into two daughter cells, the DNA in the nucleus must be duplicated, or copied, so that each new cell gets the same DNA as the original parent cell. How does this happen? Let's find out.

Counting Bases

Suppose that a DNA molecule contains 2000 nitrogen base pairs. If 30 percent of the bases are thymines, how many adenines are there in this DNA molecule?

DNA Replication

The process in which DNA molecules form exact duplicates is called **replication** (rehp-luh-KAY-shuhn). During the first step in replication, the DNA molecule separates, or unzips. As you can see in Figure 2–17, the separation takes place between the two nitrogen bases that form each rung of the DNA ladder. At the end of the first step in replication, the DNA ladder has split into two halves, or strands. In the next step, free nitrogen bases that are floating in the nucleus begin to pair up with the nitrogen bases on each strand of the DNA ladder. Remember that adenine (A) always attaches to thymine (T), and guanine (G) always attaches to cytosine (C). Once the new bases are attached, two new DNA molecules are

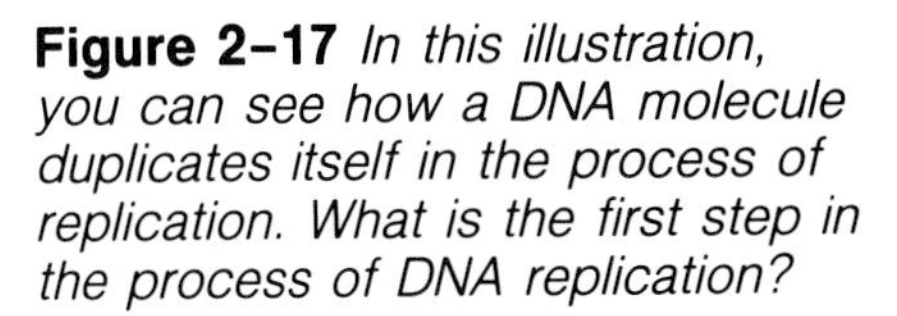

Figure 2–17 *In this illustration, you can see how a DNA molecule duplicates itself in the process of replication. What is the first step in the process of DNA replication?*

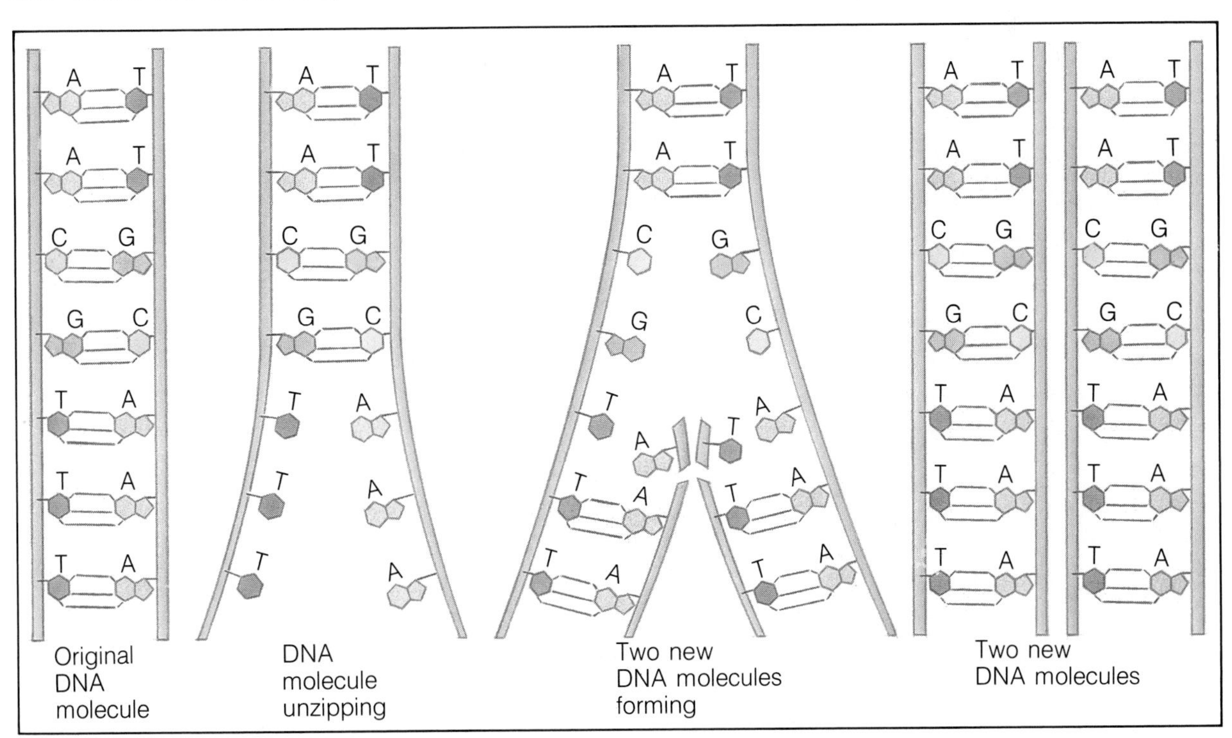

Figure 2–18 *Each body cell in this nine-banded armadillo contains all the genetic information that determines the armadillo's traits. Although you might not think so from its appearance, the armadillo is a mammal too!*

formed. Each new DNA molecule is an exact duplicate of the original DNA molecule. In other words, each new DNA molecule contains the same genetic code as the original DNA molecule and can transfer this code to a new daughter cell.

Figure 2–19 *The brightly colored stripes in these tubes are glowing bands of DNA.*

2–3 Section Review

1. What role does DNA play in heredity?
2. Describe the structure of DNA. What four nitrogen bases are found in DNA?
3. What is replication? List the steps in the process of replication.

Critical Thinking—*Making Inferences*

4. The nitrogen bases on one half of a DNA molecule are in the following order: AGTTCTCCAG. What is the order of the nitrogen bases on the other half of the molecule?

2–4 How Chromosomes Produce Proteins

Guide for Reading

Focus on this question as you read.

- *What happens during the process of protein synthesis?*

Recall that chromosomes are made up of long strands of DNA molecules. The main function of chromosomes is to control the production of substances called proteins. Proteins are the substances in the body that are necessary for building and repairing cells. Most of the chemicals that control the body's vital functions are also proteins. For example, hormones such as insulin are proteins that act as the body's chemical messengers. Enzymes such as pepsin are proteins that speed up chemical reactions in the body. These and other proteins are made in the cytoplasm of cells. The cytoplasm is the material outside the cell nucleus.

Protein Synthesis

The production of proteins is called protein synthesis. (The word synthesis means to put together.) Proteins are long molecules that are made up of chains of smaller molecules called **amino acids.** Amino acids are the building blocks of proteins. There are 20 different amino acids that join together to form protein molecules. It is the job of the DNA molecules in chromosomes to control the order in which these 20 amino acids are put together to make a protein molecule. How does DNA do this?

RNA

Protein synthesis takes place in the cytoplasm of a cell, outside the nucleus. The chromosomes containing DNA are found only inside the nucleus. The first thing that is needed in protein synthesis, therefore, is a messenger to carry the genetic code from the DNA inside the nucleus to the cytoplasm outside the nucleus. This genetic messenger is called **ribonucleic acid,** or **RNA.** RNA is similar to DNA, but with some differences.

Unlike a DNA molecule, which looks like a twisted ladder, an RNA molecule looks like only one

Essential Amino Acids

The thousands of proteins in your body are built from 20 amino acids. However, your body can make only 12 of these amino acids. You must get the other 8 from the foods you eat. These 8 amino acids are called essential amino acids. Use library references to find out which of the 20 amino acids are essential amino acids. What foods must you include in your diet to make sure you get enough of these essential amino acids? Write a report of your findings.

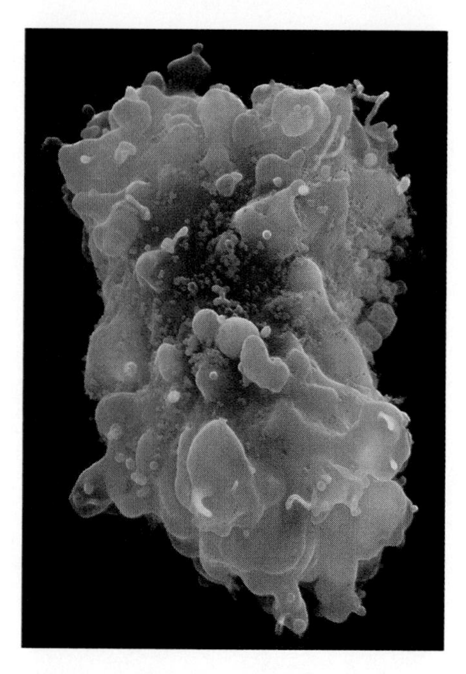
Figure 2–20 *The AIDS virus appears as small blue spots in this photograph of a human white blood cell. Where are the genes located in the AIDS virus?*

side, or strand, of the ladder. RNA also contains a different sugar molecule from the sugar found in DNA. Another difference between DNA and RNA is in their nitrogen bases. Recall that the four nitrogen bases in DNA are adenine, guanine, cytosine, and thymine. RNA also contains adenine, guanine, and cytosine. But instead of thymine, RNA contains uracil (YOOR-uh-sihl). Certain viruses (including the virus that causes AIDS) contain only RNA. In these viruses, the genes are carried on RNA molecules instead of on DNA molecules.

Translating the Genetic Code

How does an RNA molecule translate the genetic code of a DNA molecule? This process is shown in Figure 2–21. As in replication, a molecule of DNA first unzips between nitrogen base pairs. Then each strand of the DNA molecule directs the production of a complementary strand of RNA. In other words, the structure of the DNA strand determines the

Figure 2–21 *This illustration shows how an RNA molecule reads the genetic code of a DNA molecule. Which nitrogen base is found only in RNA?*

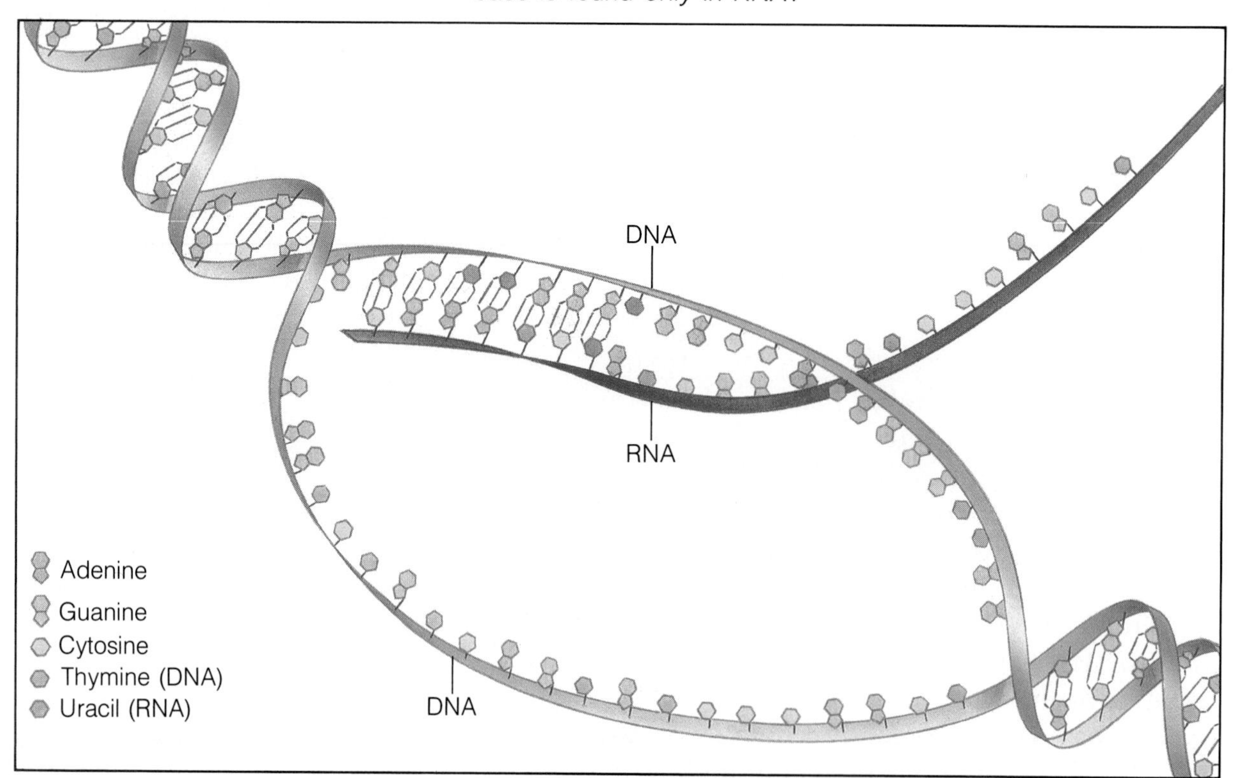

structure of the RNA strand. In the RNA strand, cytosine (C) joins with guanine (G), and adenine (A) joins with uracil (U) instead of thymine (T). Thus, the genetic information in the DNA strand is transferred to the RNA strand, which separates quickly from the DNA strand. The RNA molecule then carries the genetic information out of the nucleus and into the cytoplasm.

The genetic information that DNA transfers to RNA is in the form of a code made of three-letter code words. Each code word consists of three nitrogen bases—for example, AUG or ACA. There are 64 possible three-letter code words. Each three-letter code word specifies a particular amino acid to be added to the growing protein chain. Other code words function as "start" and "stop" signals. (AUG means "start"; UAA, UAG, and UGA mean "stop.")

The RNA molecule that carries the coded message for a specific protein does not actually produce the protein itself. Putting together the amino acids in a protein chain is the next step in protein synthesis, and it requires another form of RNA. This RNA molecule picks up the amino acids specified by the coded message and puts them into the correct order in the protein chain. A "start" code signals the beginning of the protein chain. As the amino acids line up in order, bonds form between them. When a "stop" code is reached, a complete protein molecule has been produced, and protein synthesis is complete.

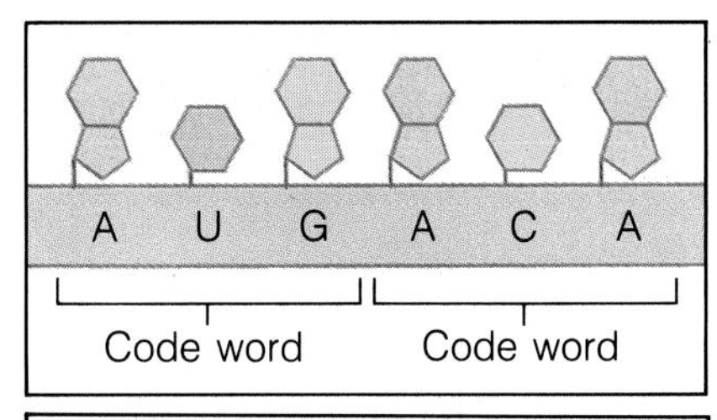

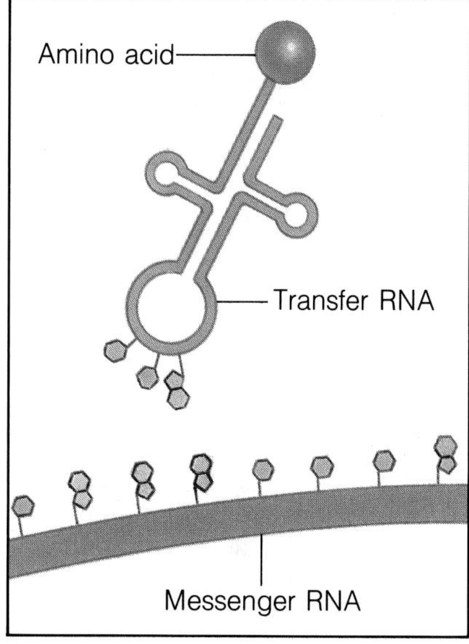

Figure 2–22 *Each combination of three nitrogen bases in a DNA molecule makes up a three-letter code word (top). These code words are copied onto a messenger RNA molecule. During protein synthesis, a transfer RNA molecule reads the code and begins putting together amino acids to make a protein chain (bottom).*

2–4 Section Review

1. What is the name for the production of proteins in the cytoplasm of a cell?
2. What is RNA? How is RNA different from DNA?
3. What are amino acids? How many amino acids are there?

Critical Thinking—*Making Inferences*

4. Why do you think code words for "start" and "stop" are necessary in protein synthesis?

Laboratory Investigation

Observing the Growth of Mutant Corn Seeds

Problem

What is the effect of a mutation on the growth of corn plants?

Materials *(per group)*

10 albino corn seeds
10 normal corn seeds
flower box
potting soil
string
tape
marking pen

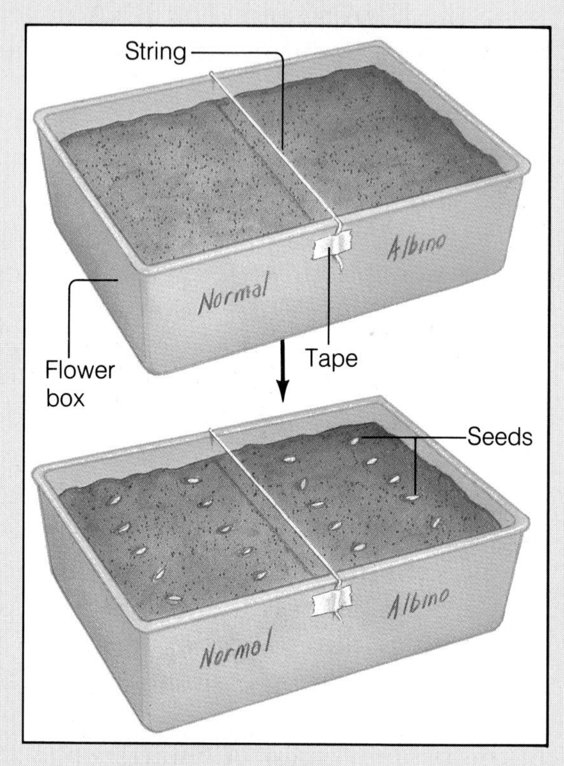

Procedure

1. Fill a flower box about three-fourths full of potting soil.
2. Use a piece of string to divide the flower box in half across the width of the box. Tape the ends of the string to the box to hold the string in place.
3. Label the right side of the box Albino. Label the left side of the box Normal.
4. On the right side of the box, plant each of the albino seeds about 1 cm below the surface of the soil. The seeds should be spaced about 1 cm apart. Water the soil.
5. On the left side of the box, plant each of the normal seeds as you did the albino seeds in step 4.
6. Place the flower box on a table near a window or on a windowsill where it will receive direct sunlight. Keep the soil moist. Observe the box every day for three weeks.

Observations

1. What was the total number of seeds that sprouted?
2. How many albino seeds sprouted? How many normal seeds?
3. What happened to the plants a week after they sprouted? Two weeks?
4. Describe the difference in appearance between the albino plants and the normal plants.

Analysis and Conclusions

1. Did the albino seeds grow as well as the normal seeds?
2. Which seeds, albino or normal, showed the mutation?
3. What effect did the mutation have on the growth of the corn plants?
4. **On Your Own** If you were a farmer, which corn plants—albino or normal—would you choose to grow for their desirable traits? Explain.

Summarizing Key Concepts

2–1 The Chromosome Theory

▲ Chromosomes are rod-shaped structures in the nucleus of an organism's cells.

▲ The chromosome theory states that chromosomes carry genes, which determine hereditary traits.

▲ The main function of chromosomes is to control the production of proteins.

▲ Meiosis is the process by which sex cells receive half the normal number of chromosomes as the parent.

▲ The X and Y chromosomes are the sex chromosomes.

▲ Human females have two X chromosomes; human males have one X chromosome and one Y chromosome.

2–2 Mutations

▲ A mutation is a sudden change in an organism caused by a change in a gene or chromosome.

▲ A mutation that takes place in a sex cell may be passed on to offspring.

▲ Mutations may be helpful or harmful.

▲ Mutagens are factors in the environment that cause mutations.

2–3 The DNA Molecule

▲ DNA stores and passes on genetic information from one generation to the next.

▲ The shape of a DNA molecule resembles a twisted ladder or spiral staircase.

▲ The four nitrogen bases in DNA are adenine, guanine, cytosine, and thymine.

▲ Replication is the process by which DNA molecules make exact copies, or duplicates, of themselves.

2–4 How Chromosomes Produce Proteins

▲ The process by which proteins are produced is called protein synthesis.

▲ Proteins are made up of chains of amino acids.

▲ RNA is necessary in order for protein synthesis to take place.

▲ Instead of thymine, RNA contains the nitrogen base uracil.

Reviewing Key Terms

Define each term in a complete sentence.

2–1 The Chromosome Theory

chromosome
meiosis
sex chromosome

2–2 Mutations

mutation
mutagen

2–3 The DNA Molecule

DNA
deoxyribonucleic acid
replication

2–4 How Chromosomes Produce Proteins

amino acid
ribonucleic acid
RNA

Chapter Review

Content Review

Multiple Choice

Choose the letter of the answer that best completes each statement.

1. The rod-shaped structures found in the nucleus of a cell are called
a. genes. c. chromosomes.
b. proteins. d. nitrogen bases.

2. The scientist who proposed the chromosome theory of heredity was
a. Watson. c. Morgan.
b. Crick. d. Sutton.

3. RNA molecules contain all of the following nitrogen bases except
a. adenine. c. cytosine.
b. guanine. d. thymine.

4. Mutations were discovered by
a. Franklin. c. De Vries.
b. Sutton. d. Morgan.

5. Proteins are made up of long chains of
a. nitrogen bases.
b. DNA molecules.
c. amino acids.
d. RNA molecules.

6. Which sex chromosomes would be found in the cells of a normal male fruit fly?
a. XX c. YY
b. XY d. XYY

7. The nitrogen base not found in DNA molecules is
a. cytosine. c. uracil.
b. thymine. d. guanine.

True or False

If the statement is true, write "true." If it is false, change the underlined word or words to make the statement true.

1. Proteins are produced through the process of <u>replication</u>.

2. Mutations that take place in body cells <u>can</u> be passed on to offspring.

3. In addition to nitrogen bases, DNA contains phosphate groups and <u>salt</u> molecules.

4. Sex cells have <u>twice</u> as many chromosomes as the parent's body cells.

5. A <u>lethal</u> mutation does not cause any obvious changes in an organism.

6. Genes are located on rod-shaped structures called <u>chromosomes</u>.

7. The nitrogen base uracil is found only in <u>DNA</u> molecules.

8. The process by which a DNA molecule makes a duplicate of itself is called <u>protein synthesis</u>.

Concept Mapping

Complete the following concept map for Section 2–1. Refer to pages E6–E7 to construct a concept map for the entire chapter.

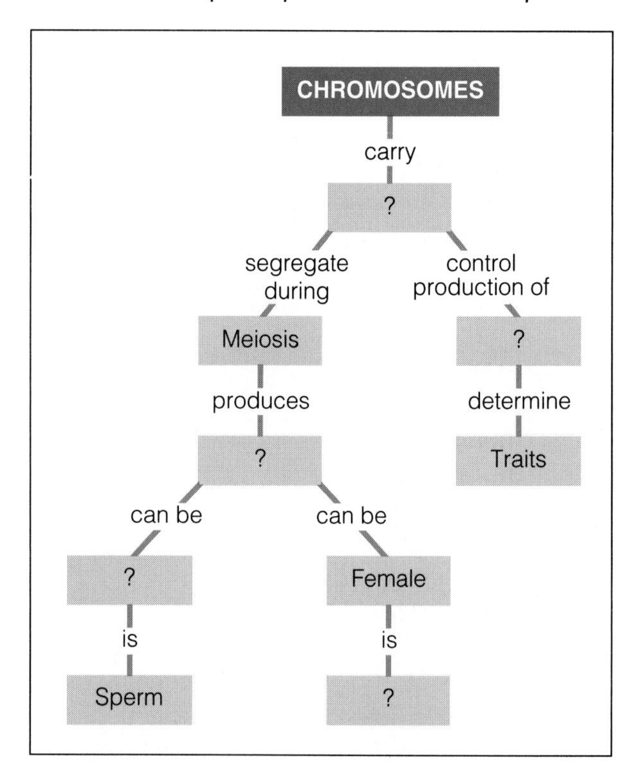

Concept Mastery

Discuss each of the following in a brief paragraph.

1. How is a molecule of RNA similar to a molecule of DNA? How are they different?
2. Describe the process of DNA replication. Why is replication necessary?
3. Explain how chromosomes determine the sex of an organism.
4. Describe the role of RNA in protein synthesis.
5. Why did Thomas Hunt Morgan choose to study fruit flies?
6. Explain how the X-ray photographs of DNA produced by Rosalind Franklin helped Watson and Crick to discover the structure of the DNA molecule.
7. Explain how mutagens can produce helpful or harmful mutations. Give specific examples.

Critical Thinking and Problem Solving

Use the skills you have developed in this chapter to answer each of the following.

1. **Applying concepts** If a woman developed skin cancer as a result of a gene mutation in the skin cells of her arm, could she pass the skin cancer gene on to her children? Explain.
2. **Relating concepts** Each body cell in a mouse contains 40 chromosomes. How many chromosomes did the mouse receive from each of its parents? How many chromosomes are present in the mouse's sex cells?
3. **Making inferences** A mutation caused one Bengal tiger in the photograph to be born white instead of orange like other Bengal tigers. Do you think this mutation would be helpful, harmful, or have no effect on the tiger? Explain. 
4. **Relating facts** A DNA molecule always contains equal amounts of adenine and thymine, as well as equal amounts of guanine and cytosine. How did this fact lead to the conclusion that adenine always joins with thymine and guanine always joins with cytosine?
5. **Applying concepts** Write out the sequence of nitrogen bases that would be copied onto a strand of RNA from the following sequence of bases in a DNA strand: TTCTTTGTTCATGAACAT. Then write out this RNA sequence as a series of three-letter code words.
6. **Making calculations** Why are there only 64 possible three-letter code words that can be formed from the four nitrogen bases?
7. **Using the writing process** Suppose you could use a mutagen to produce a specific useful mutation in a particular organism. Which organism would you choose and why? How would that trait be desirable or useful? Write a brief essay explaining your choice.

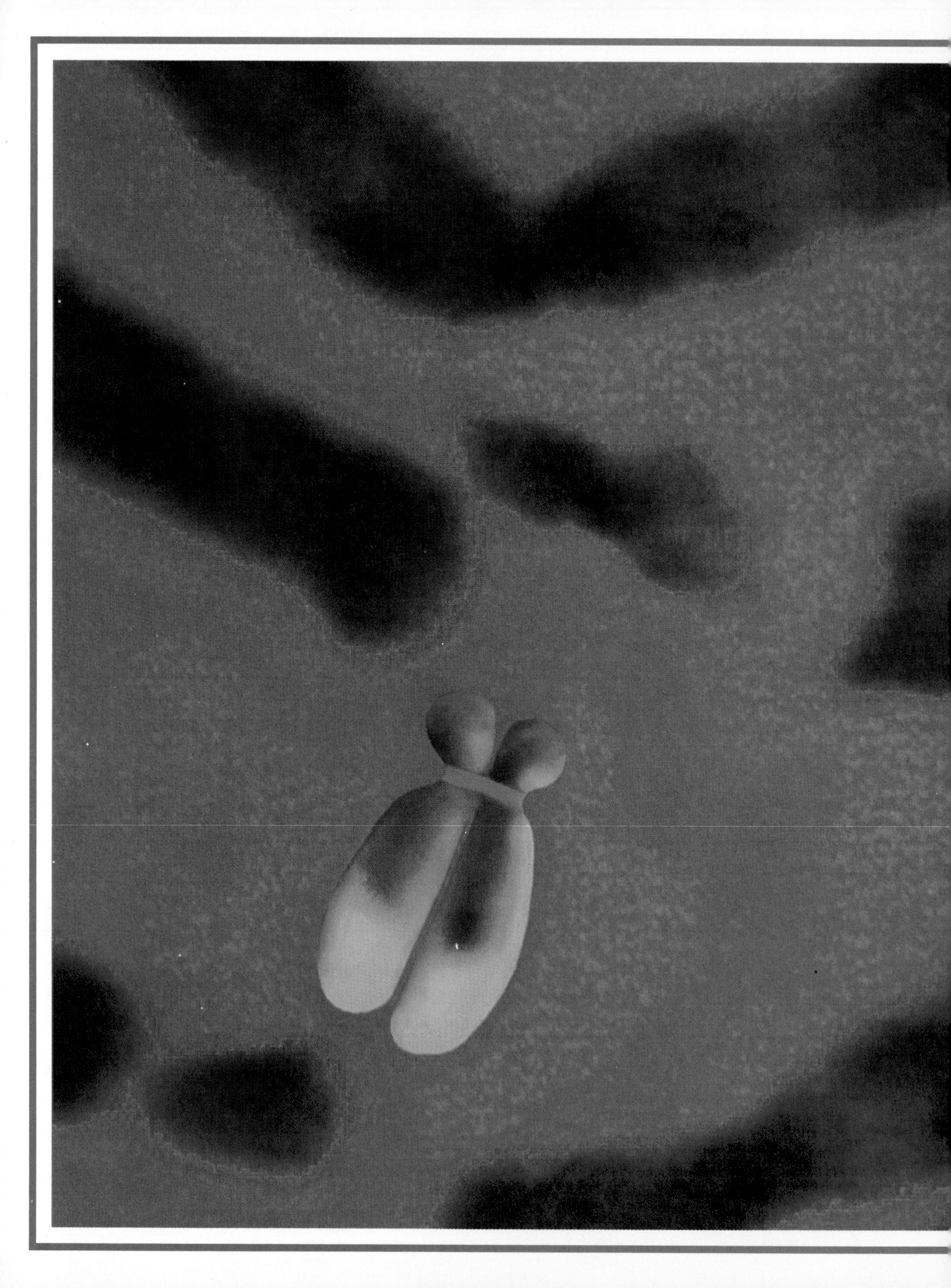

Human Genetics

Guide for Reading

After you read the following sections, you will be able to

3–1 Inheritance in Humans

- Explain how the basic principles of genetics can be applied to human heredity.

3–2 Sex-linked Traits

- Describe how sex-linked traits are inherited.

3–3 Human Genetic Disorders

- Relate the process of nondisjunction to certain human genetic disorders.
- Discuss the possibility of curing human genetic disorders.

Why are men and women different? Why aren't men more like women? Men and women have been asking these questions for thousands of years. Today, scientists know that the answers to these questions can be found in the chromosomes. Women have two X chromosomes. Men have an X chromosome and a Y chromosome. In 1959, scientists discovered that only a tiny portion of the Y chromosome determines that an unborn child will be a boy and not a girl. Since then, geneticists have been searching the Y chromosome, trying to pinpoint this "maleness" gene.

In the summer of 1990, two groups of British researchers announced that they had found what may be the "master gene" for maleness. The maleness gene seems to be a small piece of DNA on the Y chromosome that triggers the production of a protein called testosterone (tehs-TAHS-ter-ohn). Testosterone is a male sex hormone that controls the development of male characteristics, such as a deep voice and the ability to grow a beard. So a tiny piece of DNA is all that separates men from women. Or is it? Maybe science will never really be able to explain the difference between men and women!

Journal *Activity*

You and Your World Do you think men and women (or boys and girls) behave differently? In your journal, describe one or two ways in which you think a man and a woman might react differently in the same situation. If you can get two of your classmates to volunteer, you might want to test your prediction.

The maleness gene is highlighted in pink in this computer-enhanced photograph of a human Y chromosome.

Guide for Reading

Focus on this question as you read.

- *How are human traits inherited?*

Where Do Proteins Come From?, p.109

The Eyes Have It

One of the traits you may inherit from your parents is the tendency to use one eye more than the other. This is called eye dominance.

1. Hold your hand out at arm's length.

2. Point your finger at an object across the room.

3. Close your right eye and observe how far your finger seems to move. Repeat with the left eye. The eye that seems to keep your finger closer to the object is your dominant eye.

4. Combine the results for all the students in your class on a bar graph.

Are most students in your class right-eyed, left-eyed, or neither?

- Which trait—right-eye dominance or left-eye dominance—is dominant?
- Is eye dominance related to hand dominance? How could you find out?

3–1 Inheritance in Humans

Like all living things, humans are what they are because of the genes they inherit from their parents. In Chapter 2, you learned that all traits are controlled by genes, which are found on chromosomes. Each human has about 100,000 genes, located on 46 chromosomes. These 46 chromosomes are arranged in 23 pairs. Each chromosome pair has matching genes for a particular trait, such as eye color, hair color, and ear-lobe shape. Do you have brown eyes, black hair, and attached ear lobes? Or are you a blue-eyed blond with free ear lobes? Whatever your physical appearance, you inherited all of your traits from your parents.

Because you got one chromosome from each parent, you also got one gene for a particular trait from each parent. For example, you received one gene for eye color from your mother and one from your father. The way these genes combined determined whether your eyes would be brown, blue, or some other color. How do genes determine what color a person's eyes will be? Genes tell body cells what chemicals to make and how to make them. These chemicals are proteins. Special proteins called

Figure 3–1 *"Like father, like son." All the children's traits are controlled by genes, which they inherited from their parents. How do genes determine traits?*

enzymes are responsible for making the pigment, or coloring material, in your eyes.

As you can see, human genes seem to follow the same pattern of inheritance as the genes in the pea plants studied by Mendel more than a century ago. **Scientists can now apply some of the basic principles of genetics to the study of human heredity.** Today, many geneticists are in the process of mapping all 46 human chromosomes to identify the individual genes that control particular human traits.

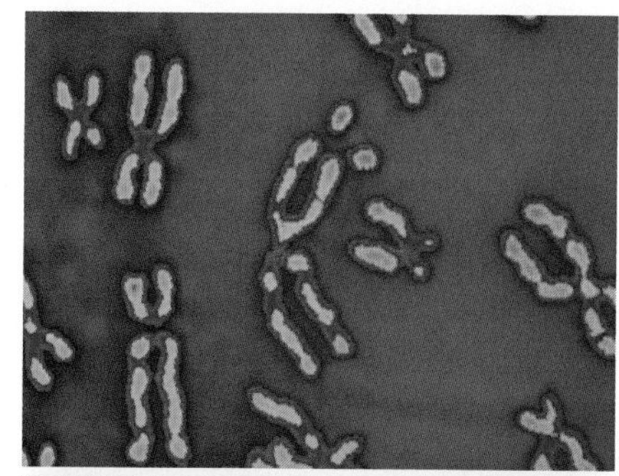

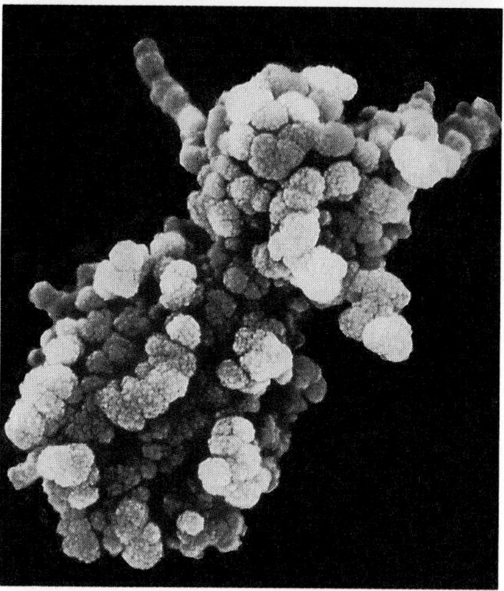

Figure 3–2 *In the photograph on the top you can see some of the 23 chromosome pairs found in human cells. One human chromosome, magnified about 20,000 times by an electron microscope, is shown in the photograph on the bottom. How many chromosomes does each human body cell contain?*

Male and Female

Recall that the X and Y chromosomes are the sex chromosomes. The sex chromosomes are the only chromosomes in which the members of a pair do not always match each other. The X chromosome is rod shaped and the Y chromosome is hook shaped. In normal human males, all the body cells have one X chromosome and one Y chromosome, or XY. Females have two matching X chromosomes, or XX. All female sex cells (eggs) contain one X chromosome. Male sex cells (sperm) may contain either an X chromosome or a Y chromosome. Sex is determined by whether an egg is fertilized by a sperm carrying an X chromosome or a Y chromosome.

How do the X and Y chromosomes determine whether a person will be male or female? Scientists have discovered that sex seems to be determined by the presence of a Y chromosome—not by the number of X chromosomes. For example, in rare cases babies may be born with an abnormal number of sex chromosomes. Babies born with only one X chromosome and no second sex chromosome (XO) are female in appearance. Babies born with two X chromosomes and one Y chromosome (XXY) are male in appearance. In both of these abnormal cases, however, the individuals will be sterile as adults; that is, they will not be able to have children.

There have been no reported cases of babies being born without an X chromosome. It seems that the presence of an X chromosome is necessary for survival. As you will see in the next section, this is because the X chromosome carries a number of genes that are needed for normal development.

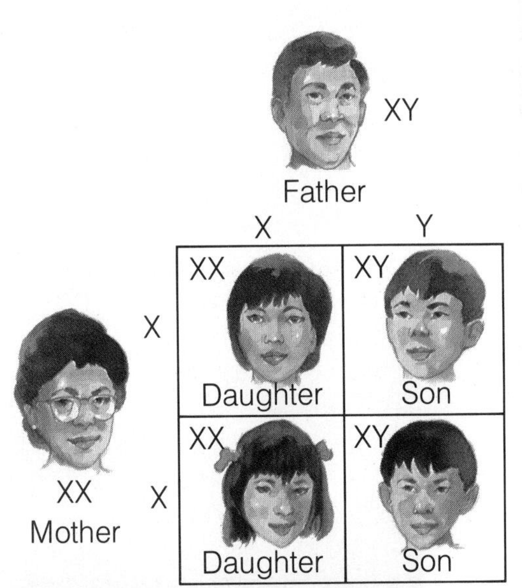

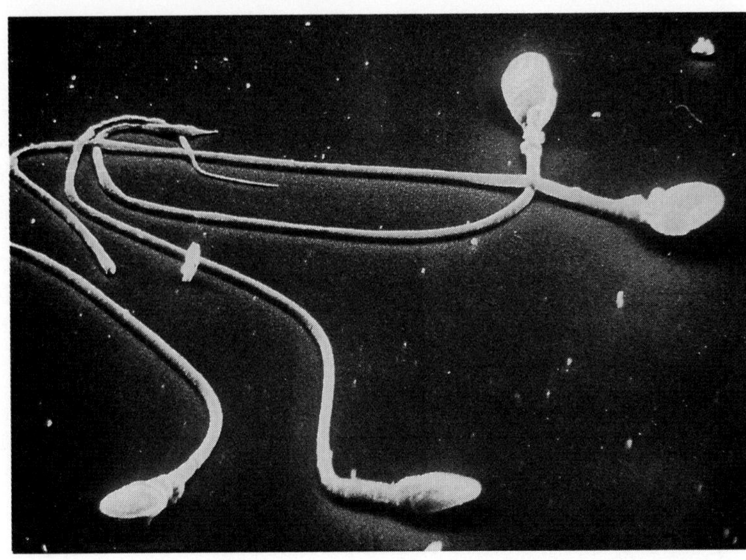

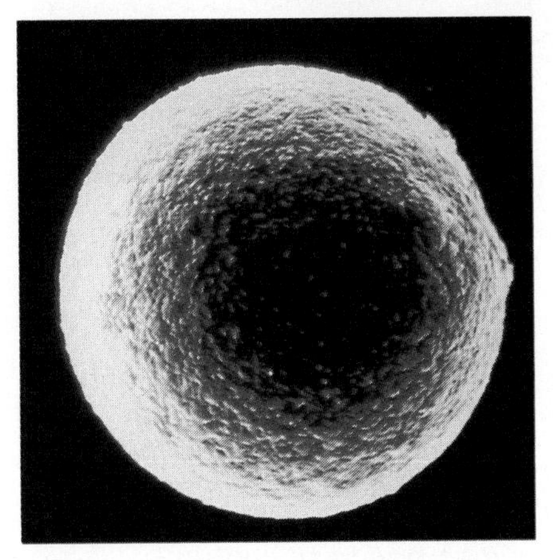

Figure 3–3 *As you can see in this Punnett square, the sex of each child is determined by whether the mother's egg is fertilized by a sperm from the father carrying an X chromosome or a Y chromosome. These photographs show a human egg (bottom) and human sperm (top).*

Multiple Alleles

In Chapter 1, you learned that particular traits in plants (such as seed color) are determined by a single pair of genes, with each gene in the pair usually being either dominant or recessive. In humans, however, some traits are not so easily determined. For example, human skin color is controlled by several genes, some of which have more than two forms.

Each form of a gene is called an **allele** (uh-LEEL). So far, you have read only about genes that have two alleles. For example, the pea plant gene for seed color has two alleles, one for yellow and one for green. In humans, however, there may be three or more alleles for a single skin-color gene. In other words, the gene has multiple alleles. Although many alleles may exist for a particular gene, each individual has only two alleles for that gene.

In humans, the inheritance of skin color can be unpredictable. Even children of the same mother and father may have different patterns of skin color. The color of human skin depends on the amount and types of brownish pigments present in the skin cells. Various combinations of the individual genes for skin color control the amount of pigments produced in the skin cells. This, in turn, results in the wide variety of skin colors in people around the world. How boring it would be if everyone in the world had the same skin color! Instead, human skin colors range from palest white to bluest black—with many, many variations in between.

In addition to skin color, the four major human blood groups, or types, are also determined by multiple alleles. The major human blood groups are called A, B, AB, and O. Scientists know that blood groups are controlled by multiple alleles because there is no way that a single pair of alleles can produce four different characteristics.

Human blood groups are determined by three alleles. Both the allele for group A blood and the allele for group B blood are dominant. In other words, they are **codominant.** When two codominant alleles are inherited, both are expressed. For example, a person who inherits an allele for group A blood from one parent and an allele for group B blood from the other parent will have group AB blood.

Make a Map

The goal of a huge research effort known as the Human Genome Project is to map, or locate, every gene on all 46 human chromosomes. Since there are about 100,000 human genes, this is truly a massive undertaking!

You are probably familiar with simpler maps, such as street maps. In this activity you will make a street map.

1. On a sheet of paper, draw the main streets and cross streets between your house and a nearby landmark. You might choose your school, the public library, or some other landmark.

2. On your map show your house, the houses of friends and relatives, local parks or playgrounds, supermarkets, and so forth.

3. Exchange street maps with a classmate. With your teacher's permission, use the map to find a specific location chosen by your classmate.

How successful were you in following your classmate's map? How are maps useful in our daily lives? What are some ways in which a map of the human genome will be helpful?

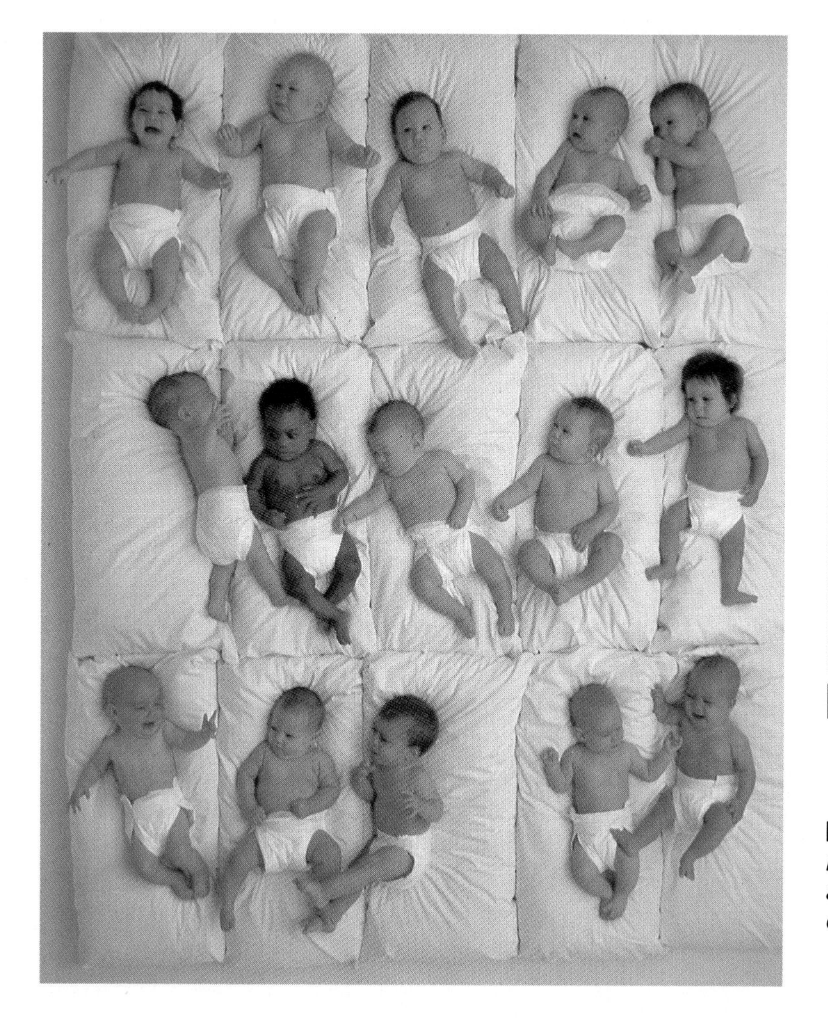

Figure 3–4 *Humans come in a rainbow of skin colors. How many alleles determine human skin color?*

BLOOD GROUP ALLELES

Blood Groups	Combination of Alleles
A	AA or AO
B	BB or BO
AB	AB
O	OO

Figure 3–5 *The photograph shows group A blood before (left) and after (right) mixing with a different blood group. Notice the clumping of blood cells that has occurred. According to the table, what two possible allele combinations might a person with group A blood have?*

The allele for group O blood, however, is recessive. So a person who inherits an allele for group O blood and an allele for group A blood will have group A blood. (The recessive O allele is masked by the dominant A allele.) Similarly, a person who inherits an allele for group O blood and an allele for group B blood will have group B blood. What two alleles must a person with group O blood have inherited? The table in Figure 3–5 shows the pattern of inheritance for human blood groups. To which blood group do you belong?

Sickle Cell Anemia

As you have already learned, each gene controls the production of a specific protein. Sometimes a mutation occurs in an inherited gene. Then the protein whose production is controlled by the mutant gene may contain an error in its structure. If the mutant gene controls the production of an important protein, such as hemoglobin, the consequences may be serious. Hemoglobin is the red pigment in red blood cells that carries oxygen. Hemoglobin that has an error in its structure may not be able to do its job properly. If so, the result may be a serious blood disorder called sickle cell anemia. Sickle cell anemia is an example of an inherited disease. It occurs when a person inherits from each parent a mutant gene for the manufacture of hemoglobin.

THE SICKLE CELL GENE People with sickle cell anemia have inherited two sickle cell genes, one from each parent. This is because the gene for normal hemoglobin (A) is codominant with the sickle cell gene (S). (Remember that when two codominant genes are inherited, both are expressed.) When each gene is present (AS), the person is said to be a carrier of the sickle cell trait. About half of a carrier's hemoglobin is normal. Carriers, therefore, show few of the harmful effects of sickle cell anemia. When both sickle cell genes (SS) are present, however, the person has sickle cell anemia and suffers all of the effects of the disorder.

CAUSE OF SICKLE CELL ANEMIA Sickle cell anemia is caused by a change in one of the nitrogen bases that make up the gene for hemoglobin. The presence of

ACTIVITY WRITING

Discovery of Sickle Cell Anemia

Sickle cell anemia was discovered and named by James Herrick, an American doctor. Use library references to find out about Dr. Herrick and how he came to discover sickle cell anemia. Write a brief report of your findings.

a different nitrogen base in the hemoglobin gene results in the substitution of a different amino acid (valine instead of glutamic acid) in the hemoglobin protein molecule. This error in the structure of the hemoglobin molecule results in the characteristic sickle-shaped red blood cells seen in the blood of people with sickle cell anemia.

DISTRIBUTION OF SICKLE CELL ANEMIA In the United States, most carriers of sickle cell anemia are African Americans. In fact, about 10 percent of African Americans carry the sickle cell trait. As many as 40 percent of the population in some parts of Africa may be sickle cell carriers. The frequency of sickle cell anemia in certain areas has to do with the relationship between sickle cell anemia and malaria. Malaria is a disease that is common in Africa and other tropical parts of the world. Malaria (like sickle cell anemia) affects the red blood cells. Scientists have found that sickle cell carriers are partially resistant to malaria. Thus the sickle cell trait probably developed as a mutation that helped people who were carriers of the trait to resist malaria.

	A	S
A	AA	AS
A	AA	AS

A = gene for normal hemoglobin
S = gene for sickle cell hemoglobin

Figure 3–6 *This Punnett square shows a cross between a person who carries a gene for sickle cell hemoglobin and a person with two genes for normal hemoglobin. What are the possible genotypes of their offspring?*

Other Inherited Diseases

In addition to sickle cell anemia, there are many other inherited diseases that result when a mutation occurs in one or more human genes. Other inherited diseases include muscular dystrophy, Huntington disease, and cystic fibrosis. In some cases, scientists have been able to identify the gene responsible for the disorder. Identifying the genes that are responsible for the disorder is the first step in finding a cure for inherited diseases.

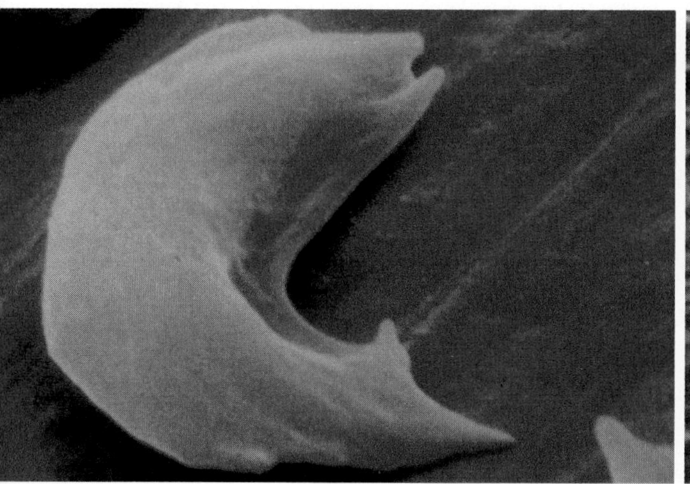

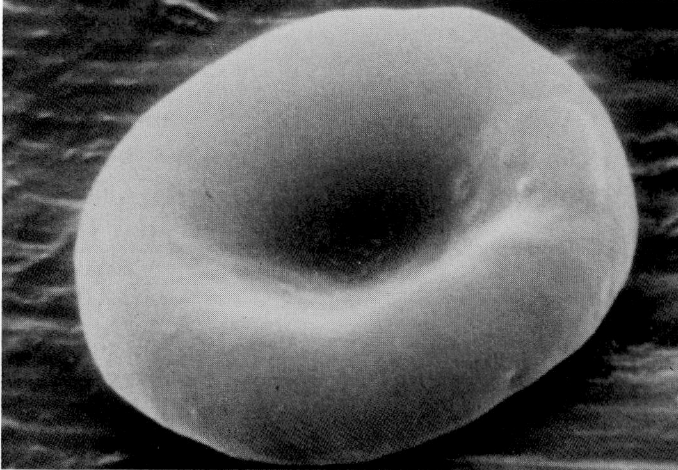

Figure 3–7 *As you can see, the shape of a red blood cell in a person who has sickle cell anemia (left) is quite different from the normal shape of a red blood cell (right). Can a person inherit sickle cell anemia from only one parent?*

Problem Solving

Pedestrian Genetics

The inhabitants of the planet Pedestria (called Pedestrians) never invented the wheel, so they walk a lot. As a result of all this walking, they have very big feet. The feet of the Pedestrians may be either red, blue, purple, or green, depending on what combination of genes for foot color they inherit from their parents. The table shows the relationship between foot color (phenotype) and gene combinations (genotype).

How many alleles are there for Pedestrian foot color? Are any of the alleles dominant? Recessive? Codominant? Suppose a male Pedestrian with blue feet (BG) marries a female Pedestrian with purple feet (RB). Is it possible for them to have a child with green feet? Why or why not? What percentage of their offspring could have each possible phenotype? Genotype?

Phenotype	Genotype
Red	RR or RG
Blue	BB or BG
Purple	RB
Green	GG

Heredity and Environment

How much of a person's appearance and behavior is controlled by heredity and how much by environment? Most people agree that physical characteristics, such as straight hair or curly hair and blue eyes or brown eyes, are inherited. But what about other physical characteristics, such as height or body weight? Height and body weight are probably also inherited traits. But people only grow to their full height and normal body weight when they receive a proper diet, and diet is a factor that is determined by the environment.

The roles of heredity and environment are easier to evaluate in organisms that have the same or similar genetic makeup. To analyze the effects of heredity and environment in humans, scientists often study identical twins. Identical twins make up only 0.5 percent of the total population. Unlike fraternal twins, who develop from two different fertilized eggs, identical twins come from the same fertilized egg and are usually genetically identical. Therefore, most differences between identical twins probably

Figure 3–8 *Because identical twins develop from the division of one fertilized egg, they have the exact same chromosomes and are always the same sex. Fraternal twins, who develop from two different fertilized eggs, are not necessarily the same sex.*

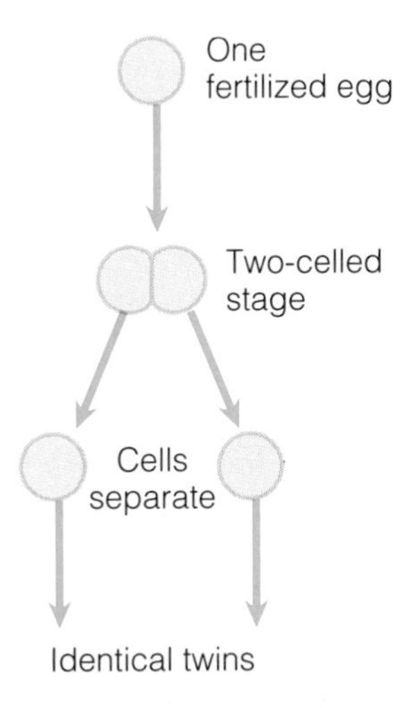

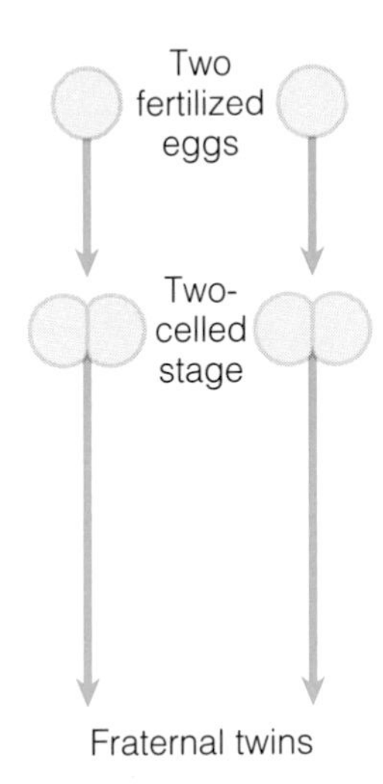

are the result of environmental factors, not heredity. Studies of identical twins who were separated at birth or at an early age suggest that more of human behavior may be inherited than was previously thought.

3–1 Section Review

1. Describe the inheritance of a human trait using a specific example.
2. What are multiple alleles? Give an example of a human trait controlled by multiple alleles.
3. What is an inherited disease? Give at least two examples.

Critical Thinking—*Relating Concepts*

4. A woman with group O blood marries a man with group B blood. Can they have a child with group A blood? Why or why not?

Guide for Reading

Focus on these questions as you read.

- *What are sex-linked traits?*
- *What are some human sex-linked traits?*

3–2 Sex-linked Traits

Some human traits occur more often in one sex than in the other. Usually, the genes for these traits are carried on the X chromosome, which is a sex chromosome. **Traits that are carried on the X chromosome are called sex-linked traits because they are passed from parent to child on a sex chromosome.** Unlike X chromosomes, Y chromosomes carry few, if any, additional genes. (The maleness gene is one of the few genes carried on the Y chromosome.) So any gene—even a recessive one—carried on an X chromosome will produce a trait in a male who inherits the gene. There is no matching gene on the Y chromosome to mask, or hide, the gene on the X chromosome. The situation is not the same for a female, however. Do you know why? Because a female has two X chromosomes, a recessive gene on one X chromosome can be masked, or hidden, by a dominant gene on the other X chromosome. As a result, females are less likely than males to inherit **sex-linked traits.**

Hemophilia

An example of a disorder caused by a sex-linked recessive trait is hemophilia (hee-moh-FIHL-ee-uh). Hemophilia is an inherited disease in which the blood clots abnormally slowly or not at all. Hemophilia is also called "bleeder's disease." For a person who has hemophilia, even a small cut or bruise can be extremely dangerous. People who have hemophilia often have to receive regular blood transfusions. In the 1980s, people who had hemophilia and others who received blood transfusions (for example, during surgery) were in danger of contracting AIDS from contaminated blood. Now, however, the blood supply is routinely screened for the presence of the AIDS virus.

Figure 3–11 on page 68 is an example of a chart called a pedigree. A pedigree (often called a "family tree") shows the relationships among the individuals in a family. A pedigree such as the one in Figure 3–11 can also be used to trace the inheritance of a particular trait in a family. The trait recorded in a pedigree could be an ordinary trait, such as hair color, or a disorder, such as hemophilia. The human pedigree in Figure 3–11 traces the pattern of inheritance of hemophilia in the royal families of Europe beginning with Queen Victoria of England. By studying

Figure 3–9 *Ryan White, who had hemophilia, contracted AIDS from a transfusion with contaminated blood. Before he died in 1990, he fought for the rights of others with AIDS. Ryan White's courageous battle against discrimination has been an inspiration to many people.*

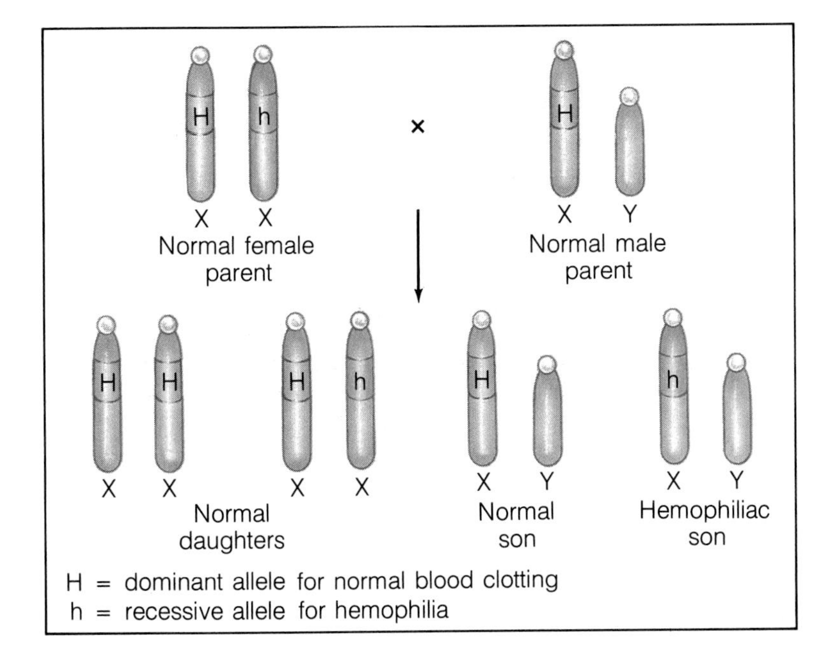

Figure 3–10 *As shown in this illustration, there is a 25 percent chance that a female who carries a gene for hemophilia and a normal male will have a son with hemophilia. Why is hemophilia called a sex-linked trait?*

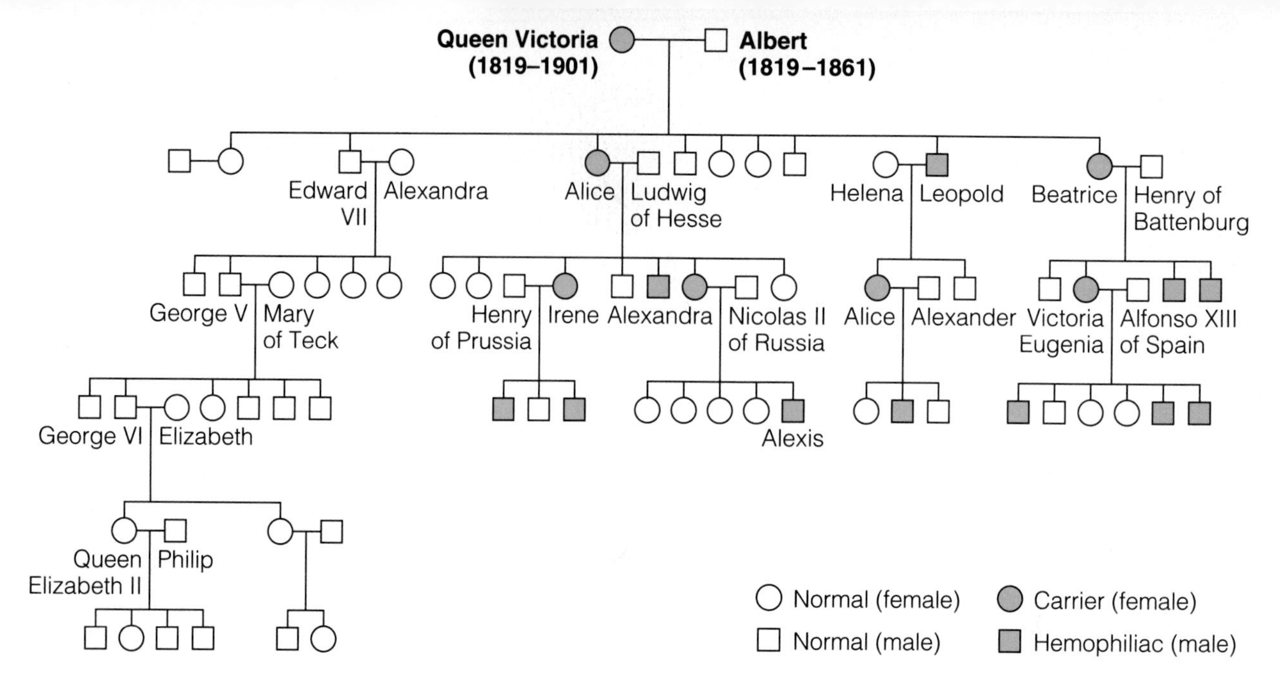

Figure 3–11 *This pedigree shows how hemophilia spread in the family of Queen Victoria of England, who was a carrier. How many of Victoria's sons had hemophilia? How many grandsons?*

the pattern of inheritance revealed in a pedigree, it is possible to determine whether a trait is dominant or recessive, as well as whether it is sex linked.

Colorblindness

Colorblindness is another sex-linked recessive trait. A person who is colorblind cannot see the difference between certain colors, such as red and green. Difficulty in distinguishing between the colors red and green is the most common type of colorblindness. More males than females are colorblind. A colorblind female must inherit two recessive genes for colorblindness, one from each parent. But a colorblind male needs to inherit only one recessive gene. Why is this so? Remember that males do not have a matching gene on the Y chromosome that could mask the recessive gene on the X chromosome.

X^C X^c Normal female — X^C Y Normal male

Sex cells: X^C, X^c, X^C, Y

	X^C	X^c
X^C	$X^C X^C$ Normal female	$X^C X^c$ Normal female
Y	X^C Y Normal male	X^c Y Colorblind male

Figure 3–12 *Colorblindness is a sex-linked trait. Why are very few females colorblind?*

Male-Pattern Baldness

Some traits that seem to be sex linked are actually not caused by genes on the X chromosome. For example, baldness is much more common in men than in women. So you might think that baldness is a sex-linked trait. However, male-pattern baldness is a sex-influenced trait. A sex-influenced trait is a trait that is expressed differently in males than it is in

Figure 3–13 *In male-pattern baldness, hair loss results from the combination of an inherited trait and the effects of male hormones.*

females. It is called male-pattern baldness because men who inherit one gene for normal hair and one gene for baldness tend to be bald, whereas women do not. Scientists are not sure how a person's sex influences the expression of certain genes, but they think that male sex hormones may play a role.

3–2 Section Review

1. What are sex-linked traits? Describe two human sex-linked traits.
2. What is a pedigree? Why is it useful?
3. Why are males more likely than females to inherit sex-linked recessive traits?
4. What is the difference between a sex-linked trait and a sex-influenced trait?

Connection—*Mathematics*

5. In a typical population, about 8 percent of the men are colorblind, but only about 1 percent of the women are colorblind. In a city of 250,000 people, about how many men would you expect to be colorblind? How many women?

Make a Family Tree

How are traits inherited in your family? In this activity you will draw a family tree to find out.

1. Choose a specific trait that you can trace through your family. Some traits you might include are hair color, left- or right-handedness, and attached or unattached earlobes.

2. Interview members of your family to trace the inheritance of the trait.

3. Draw a family tree. Use squares to represent male members of your family and circles for females. Include as many generations as you can.

Guide for Reading

Focus on this question as you read.

- *How can nondisjunction result in a genetic disorder?*

3–3 Human Genetic Disorders

In Chapter 2, meiosis was described as the process in which sex cells are formed. During meiosis, chromosome pairs usually separate. But in rare cases, a chromosome pair may remain joined during meiosis. This failure of a chromosome pair to separate during meiosis is known as **nondisjunction** (nahn-dihs-JUHNG-shuhn). **As a result of nondisjunction, body cells receive either more chromosomes or fewer chromosomes than normal. An abnormal number of chromosomes may result in certain genetic disorders.** Some of these disorders can be detected by examining a person's chromosomes.

Down Syndrome

Figure 3–15 is an example of a human **karyotype** (KAR-ee-uh-tighp). A karyotype shows the size, number, and shape of all the chromosomes in an organism. Look carefully at this human karyotype. How many chromosomes do you see? If you counted 47, you are correct. You have learned that humans usually have 46 chromosomes, or 23 pairs. In this karyotype, the extra chromosome is found in what would normally be the twenty-first chromosome pair. The presence of such a group of three chromosomes is called trisomy (trigh-SOH-mee). When a person has an extra chromosome in the twenty-first pair, a condition called trisomy-21 results. Trisomy-21 is also known as Down syndrome. People with Down syndrome may have various physical problems and some degree of mental retardation. However, many people with Down syndrome lead normal, active lives and often make valuable contributions to society.

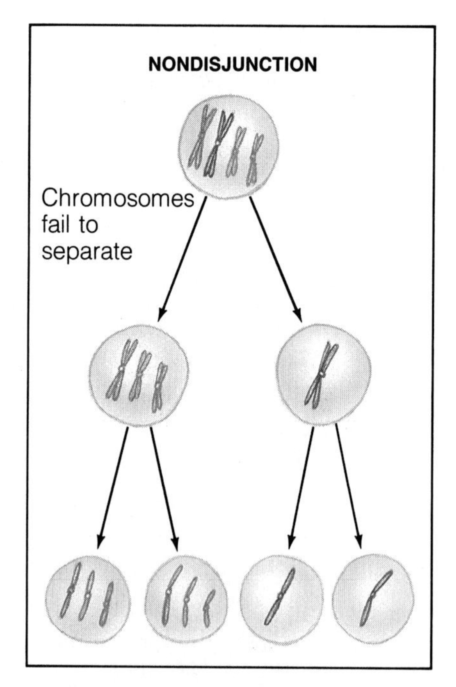

Figure 3–14 *Nondisjunction results when chromosomes fail to separate during meiosis. What is one genetic disorder that results from nondisjunction?*

Detecting Genetic Disorders

Is there a way of knowing before a child is born whether he or she will have Down syndrome or another inherited disorder? Fortunately, the answer is yes. One method of diagnosing a genetic disorder such as Down syndrome is called **amniocentesis** (am-nee-oh-sehn-TEE-sihs). Amniocentesis involves

the removal of a small amount of fluid from the sac that surrounds a baby while it is still inside its mother's body. This fluid contains some of the baby's cells. Using a microscope, doctors can examine the chromosomes in these cells. In this way, doctors can discover whether or not an unborn child has Down syndrome. A new test that gives faster results than amniocentesis is now sometimes used as an alternative. This test requires the removal of cells from the membrane surrounding the developing baby.

Different tests can reveal the presence of other inherited disorders in addition to Down syndrome. Some tests to detect a genetic disorder can make it possible to treat the disorder before birth. Other tests can be performed immediately after birth so that treatment for a detected disorder can begin as soon as possible. Parents who are concerned that they might pass a genetic disorder on to their children should consult a genetic counselor.

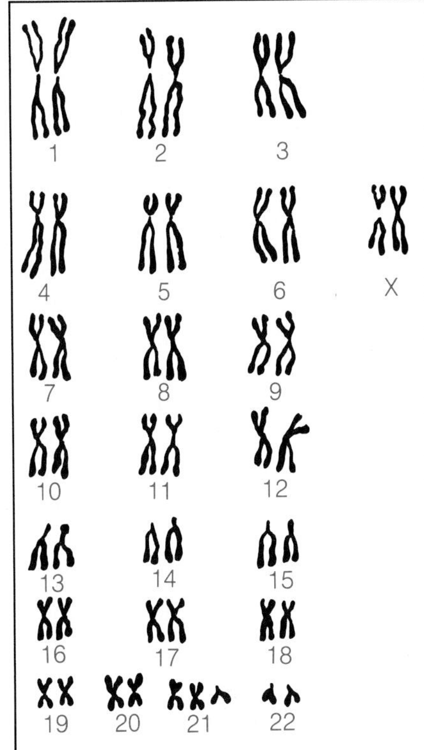

Figure 3–15 *Down syndrome is a genetic disorder in which all the body cells have an extra twenty-first chromosome. Although people with Down syndrome are mentally and physically challenged, many lead full, active, and productive lives.*

Curing Genetic Disorders

At present, there are no cures for genetic disorders. However, doctors and scientists are working to develop possible cures. In the early 1980s, scientists discovered a chemical that could slightly change the structure of the gene that causes sickle cell anemia. Changing the structure of the sickle cell gene could help people with sickle cell anemia to produce larger amounts of normal hemoglobin. In 1982, doctors at three different hospitals in the United States gave the chemical to three patients with sickle cell anemia.

Incidence of Down Syndrome

In the United States, one baby in 800 is born with Down syndrome. If 2400 babies are born in a town in one year, how many of the babies might be born with Down syndrome?

CAREERS

Genetic Counselor

Human genetics is one of the fastest changing fields in medical science today. Although many doctors specialize in diagnosing and treating genetic disorders, they often do not have the time to discuss specific genetic disorders with patients and family members. Instead, a **genetic counselor** may provide this information. A genetic counselor talks with parents who are concerned that they might be carrying genes for a disorder that they could pass on to their children.

If you are interested in this fascinating and rewarding career, you can write for more information to the March of Dimes, Birth Defects Foundation, 1275 Mamaroneck Avenue, White Plains, NY 10605.

Every patient's condition improved. But the doctors cautioned that more tests must be performed before the chemical could be tried with large groups of patients. Then, if the tests are successful, sickle cell anemia would be curable.

In 1989, scientists identified the gene responsible for cystic fibrosis. Cystic fibrosis is the most common fatal inherited disease in the United States. Less than one year later, the scientists removed cells from cystic fibrosis patients and replaced the defective genes in these cells with normal genes. As a result of these laboratory experiments, the cystic fibrosis cells were "cured." It is a long way from being successful in the laboratory to curing an inherited disease in a person. Nevertheless, the first step has been taken toward such a cure for cystic fibrosis.

Ethical Questions

Probably no one would deny that the ability to detect genetic disorders is a good thing. But this ability also raises some very important questions for society as a whole. Should we allow children to be born with serious or fatal genetic disorders? When such children are born, who should be responsible for the cost of the expensive treatment they require? Can insurance companies refuse health insurance or life insurance to people who are carriers of certain genetic disorders? These and other similar questions are difficult to answer—if, in fact, they can be answered at all. But any attempt to find a solution will involve more than science—it will involve the human spirit.

3–3 Section Review

1. What is nondisjunction? How does it cause a genetic disorder?
2. What is another name for Down syndrome? Why is this other name appropriate?
3. What is amniocentesis?

Critical Thinking—*Applying Concepts*

4. Could a karyotype be used to detect a genetic disorder? Explain.

CONNECTIONS

Hemophilia in History

Queen Victoria of England had a son and three grandsons with hemophilia. Victoria and at least two of her daughters and four of her granddaughters were carriers of the disease. That is, they carried the gene for hemophilia on one X chromosome. They did not have the disease because they carried a normal gene on the other X chromosome. However, they could pass the disease on to their offspring. Hemophilia spread throughout the royal families of Europe as Victoria's descendants passed the hemophilia gene on to their offspring.

Princess Alexandra, one of Queen Victoria's granddaughters, married the Russian czar Nicholas II. Alexandra was a carrier of hemophilia. She passed the disease on to her son, the czarevitch Alexis, who was the heir to the throne. Although Alexandra had no experience in ruling, she greatly influenced the actions of her husband, the czar. Unfortunately, she often made bad decisions based on her concern for her son. The monk Rasputin had convinced Alexandra that he could cure Alexis. As a result of his control over Alexandra, Rasputin was able to direct the czar's actions as well. The people's anger at Rasputin's evil influence over the royal family may have played some part in the Russian Revolution of 1917, in which the czar was overthrown.

Laboratory Investigation

Reading a Human Pedigree

Problem

How can you use a human pedigree to trace the inheritance of sickle cell anemia through several generations of a family?

Materials *(per student)*

paper
pencil

Procedure

1. Study the following key for the symbols used on a human pedigree.
2. Study the pedigree shown here. This pedigree traces the pattern of inheritance of sickle cell anemia in several generations of a single family.

Observations

1. How many generations are shown on the pedigree?
2. Which parent in the first generation had sickle cell anemia?
3. How many children were born in the second generation?
4. How many of these children are carriers of sickle cell anemia?
5. How many children in the third generation have sickle cell anemia? How many are carriers?

Analysis and Conclusions

1. Is sickle cell anemia a sex-linked trait? How can you tell?
2. Is the gene for sickle cell anemia more likely to be dominant or recessive? Explain your answer.
3. **On Your Own** How could a genetic counselor use a pedigree to advise parents who are worried about passing on an inherited disorder to their children?

: Male
: Female
: Parents
: Has trait
: Carrier

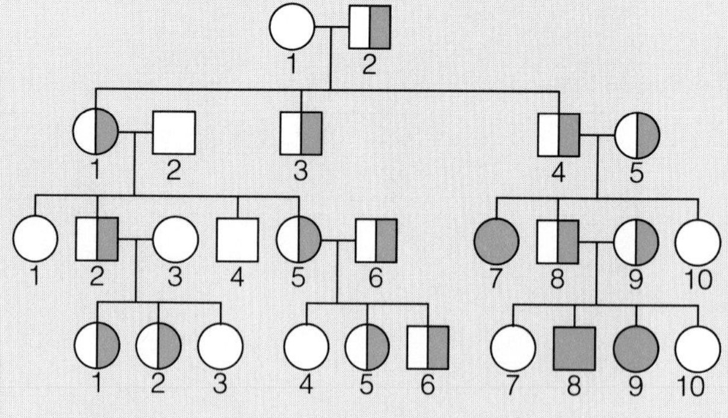

SICKLE CELL ANEMIA PEDIGREE

Study Guide

Summarizing Key Concepts

3–1 Inheritance in Humans

▲ Scientists can now apply some of the basic principles of genetics to the study of human heredity.

▲ Humans have 46 chromosomes arranged in 23 pairs.

▲ Sex is determined by the presence of a Y chromosome. Normal human males are XY; normal human females are XX.

▲ Each member of a gene pair that controls a specific trait is an allele.

▲ Some human traits, such as blood group and skin color, are controlled by multiple alleles.

▲ Sickle cell anemia is an example of an inherited disease caused by a gene mutation.

▲ A person's characteristics are determined by both heredity and environment.

3–2 Sex-linked Traits

▲ Traits that are carried on the X chromosome are called sex-linked traits.

▲ Males are more likely than females to inherit sex-linked traits.

▲ Two examples of inherited disorders caused by sex-linked recessive traits are hemophilia and colorblindness.

▲ A sex-influenced trait, such as male-pattern baldness, is a trait that is expressed differently in males than it is in females.

3–3 Human Genetic Disorders

▲ Down syndrome is a human genetic disorder caused by nondisjunction, or the failure of a chromosome pair to separate during meiosis.

▲ A karyotype shows the size, number, and shape of all the chromosomes in an organism.

▲ Some genetic disorders can be detected by means of tests such as amniocentesis.

▲ At present, there are no cures for human genetic disorders.

Reviewing Key Terms

Define each term in a complete sentence.

3–1 Inheritance in Humans

allele
codominant

3–2 Sex-linked Traits

sex-linked trait

3–3 Human Genetic Disorders

nondisjunction
karyotype
amniocentesis

Chapter Review

Content Review

Multiple Choice

Choose the letter of the answer that best completes each statement.

1. How many pairs of chromosomes do human body cells normally contain?
a. 46
b. 47
c. 23
d. 21

2. An example of a human trait that is controlled by multiple alleles is
a. eye color.
b. blood group.
c. maleness.
d. height.

3. "Bleeder's disease" is more accurately known as
a. sickle cell anemia.
b. hemophilia.
c. Down syndrome.
d. cystic fibrosis.

4. The genetic disorder caused by the wrong amino acid in hemoglobin is
a. sickle cell anemia.
b. hemophilia.
c. Down syndrome.
d. cystic fibrosis.

5. The human blood group that is determined by recessive alleles is
a. A.
b. B.
c. AB.
d. O.

6. A baby girl who inherits an A allele from her mother and an O allele from her father will have blood group
a. A.
b. B.
c. AB.
d. O.

7. A sex-linked trait is carried on
a. the X chromosome only.
b. the Y chromosome only.
c. both the X and the Y chromosome.
d. either the X or the Y chromosome.

8. Two human sex-linked disorders are hemophilia and
a. sickle cell anemia.
b. colorblindness.
c. Down syndrome.
d. male-pattern baldness.

True or False

If the statement is true, write "true." If it is false, change the underlined word or words to make the statement true.

1. The sex of a person is determined by the presence of <u>an X</u> chromosome.

2. A <u>karyotype</u> can also be called a family tree.

3. At present, it <u>is not</u> possible to cure genetic disorders.

4. Sex-linked traits are carried on the <u>Y</u> chromosome.

5. Colorblindness is <u>more</u> common in women than in men.

6. Sickle cell anemia is determined by <u>recessive</u> genes.

7. Male-pattern baldness is an example of a <u>sex-linked</u> trait.

Concept Mapping

Complete the following concept map for Section 3–1. Refer to pages E6–E7 to construct a concept map for the entire chapter.

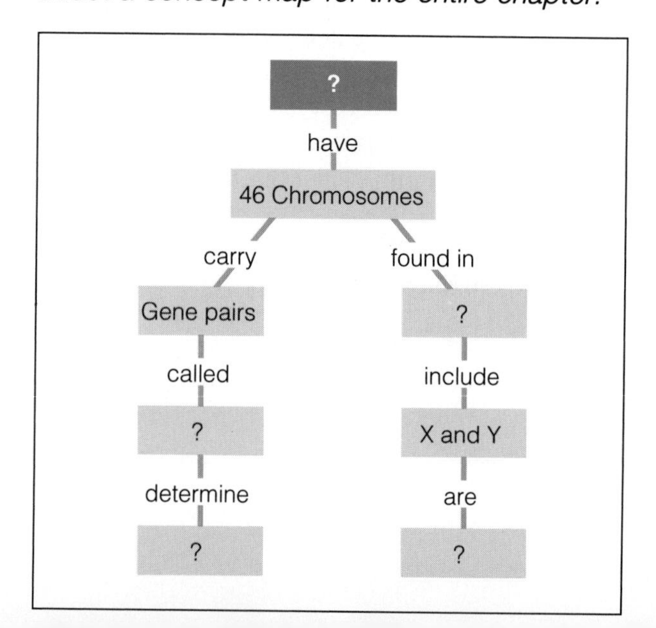

Concept Mastery

Discuss each of the following in a brief paragraph.

1. Explain how human blood groups are inherited.
2. Compare the two inherited blood diseases sickle cell anemia and hemophilia.
3. Explain the difference between a sex-linked trait and a sex-influenced trait. Use examples to describe the inheritance of a sex-linked trait and a sex-influenced trait.
4. How do the genes you inherited from your mother and father determine the color of your eyes?
5. Does colorblindness occur more frequently in men or in women? Explain.
6. What is nondisjunction? How can nondisjunction cause a genetic disorder?
7. Describe the process of amniocentesis. How is amniocentesis used to detect a disorder such as Down syndrome?
8. Why is sickle cell anemia in the United States most common among people of African American descent?

Critical Thinking and Problem Solving

Use the skills you have developed in this chapter to answer each of the following.

1. **Making predictions** Mrs. Smith has the genotype AO for blood group. Mr. Smith has the genotype BB. Predict the possible blood groups of their children. Draw a Punnett square to help you make your prediction.
2. **Relating concepts** A common form of an inherited disease called muscular dystrophy is caused by a mutation in a gene that is carried on the X chromosome. Is muscular dystrophy an example of a sex-linked disorder? How do you know?
3. **Applying concepts** A man who is colorblind marries a woman who is a carrier of the gene for colorblindness. What is the probability of their having a son who is colorblind? A daughter?
4. **Interpreting a diagram** The Punnett square shown here illustrates sex determination in humans. What are the parents' chances of having a son? A daughter? Which parent determines the sex of a child? Explain.
5. **Using the writing process** As the science reporter for your school newspaper, you have been assigned to interview a doctor who is doing research to try to find a cure for a genetic disorder. Write out a list of questions that you would like to ask the doctor.

XX Mother × XY Father

	X	Y
X	XX Daughter	XY Son
X	XX Daughter	XY Son

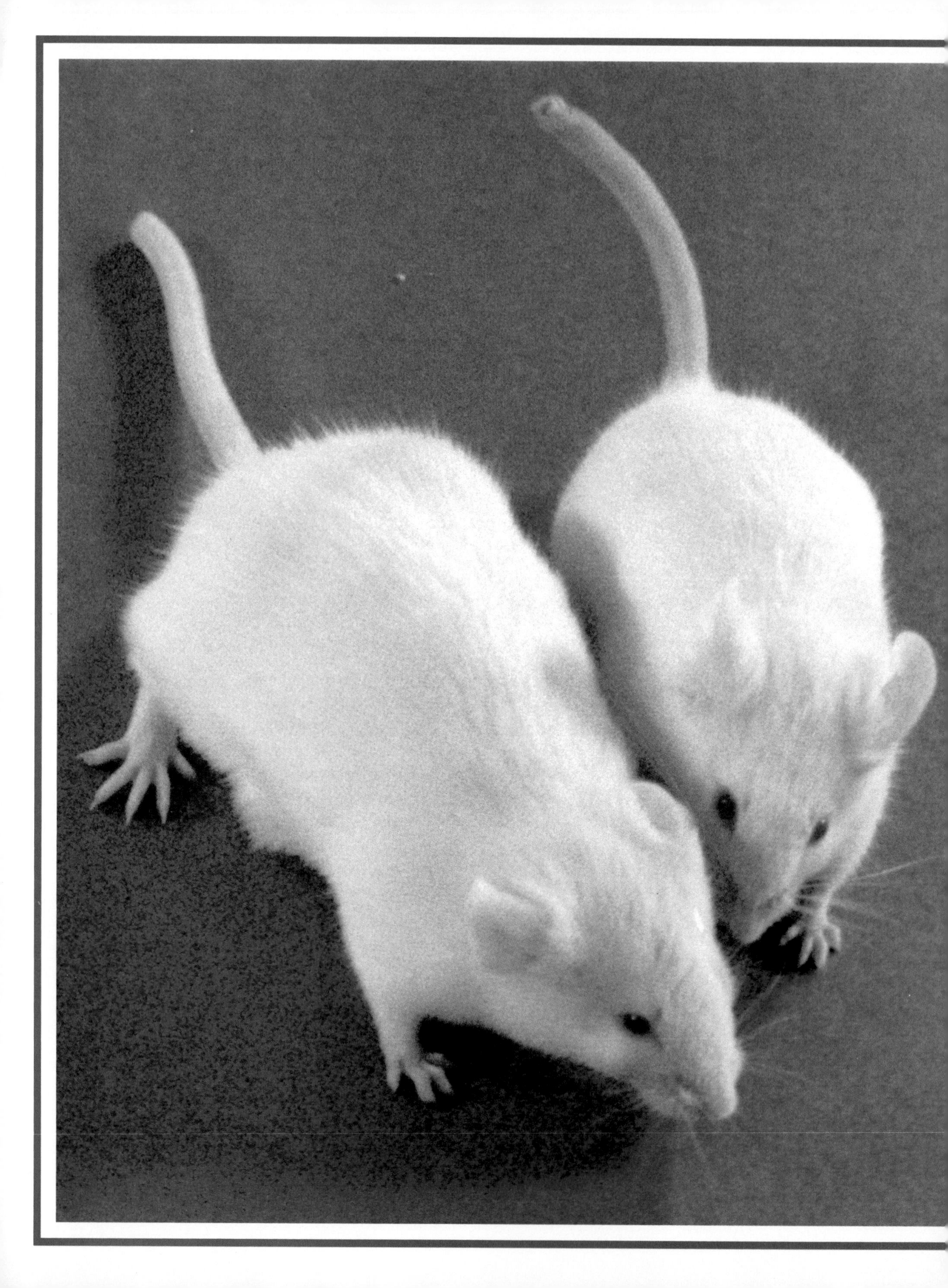

Applied Genetics

Guide for Reading

After you read the following sections, you will be able to

4–1 Plant and Animal Breeding

- Define selective breeding.
- Explain the difference between hybridization and inbreeding.

4–2 Genetic Engineering

- Describe some examples of genetic engineering.
- Explain how recombinant DNA is produced.
- Describe some applications of genetic engineering.

Do you know how to make a supermouse? Biologists at the University of Pennsylvania do. To make a supermouse, they first take a fertilized egg from an ordinary mouse. Then they place the egg under a microscope that magnifies the egg about 400 times. Now comes the really tricky part. Very carefully, the scientists inject a clear liquid containing a new gene into the mouse egg. How does this procedure produce a supermouse?

The answer to this question is hidden in the new gene. Remember that genes control the production of specific proteins. The new gene that is injected into the fertilized mouse egg controls the production of a protein called rat growth hormone. Inside the fertilized mouse egg, the transplanted new gene causes the developing mouse to produce rat growth hormone. As the mouse develops, the hormone causes it to grow to twice its normal size. It grows into a supermouse!

In this chapter you will learn how humans have used genetics to produce more nutritious crop plants and stronger, healthier farm animals. You will also find out how scientists are learning to apply the principles of genetics in medicine and agriculture.

Journal *Activity*

You and Your World What do you know about the controversy over the use of genetic engineering in agriculture and medicine? In your journal, describe what you know about genetic engineering and explain why it is controversial.

These two mice from the same litter illustrate the effect of growth hormone. The mouse on the left, which contains the growth hormone gene, is almost twice the size and weight of the normal mouse on the right.

Guide for Reading

Focus on this question as you read.

▶ *What are hybridization and inbreeding?*

Activity Bank

How Can You Grow a Plant From a Cutting?, p.110

4–1 Plant and Animal Breeding

More than 12,000 years ago, people living in the part of the world now called Iraq discovered that wild wheat could be used as food. Through a process of trial and error, these early farmers were able to select and grow wheat that had larger and more nutritious grains than the original wild wheat. People have been breeding plants and animals to produce certain desired traits ever since. This process is called **selective breeding.** Selective breeding is the crossing of plants or animals that have desirable characteristics to produce offspring with those desirable characteristics. Through selective breeding, modern plant and animal breeders are able to produce organisms that are larger in size, provide more food, or are resistant to certain diseases. For example, leaner cattle produce low-fat beef that is more healthful than the beef from fatter cattle. What other examples of selective breeding are you familiar with?

Figure 4–1 *Selective plant and animal breeding has resulted in modern strains of nutritious wheat and disease-resistant cattle.*

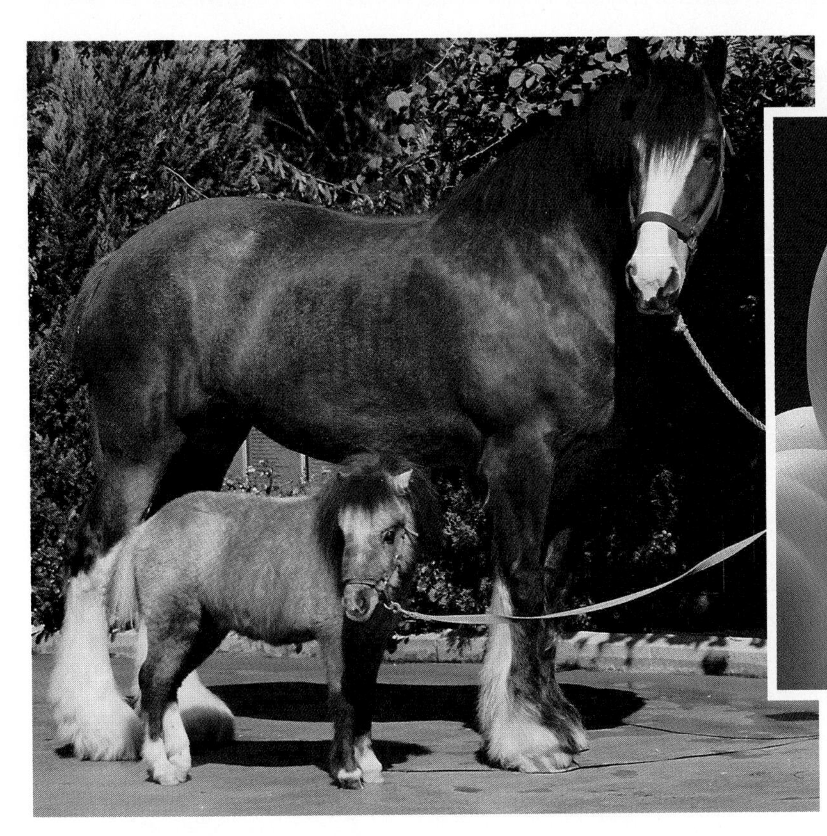

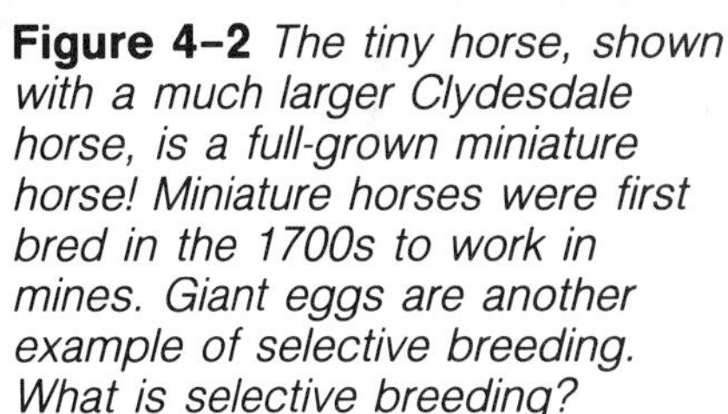

Figure 4–2 *The tiny horse, shown with a much larger Clydesdale horse, is a full-grown miniature horse! Miniature horses were first bred in the 1700s to work in mines. Giant eggs are another example of selective breeding. What is selective breeding?*

Hybridization

Sometimes breeders produce desired traits in the offspring by combining two or more different traits from the parents. To do this, breeders use a technique called **hybridization** (high-brihd-ih-ZAY-shuhn). **Hybridization is the crossing of two genetically different but related species of organisms.** When the organisms are crossed, a hybrid is produced. (Recall from Chapter 1 that a hybrid is an organism that has two different genes for a particular trait.) A hybrid organism is bred to have the best traits of both parents. For example, a mule is a hybrid that combines the traits of two different species, horses and donkeys. A mule is the offspring of a female horse and a male donkey.

Some hybrids are produced naturally. Ancient wild wheat, for example, was a hybrid that formed naturally from the crossing of one species of wild wheat with a species of wild goat grass. The result was a wheat plant with nutritious grains that could be made into bread. Early farmers were able to preserve this new hybrid wheat by selecting some of the best grains and planting them for the next harvest.

Luther Burbank

The American plant breeder Luther Burbank produced hundreds of new plant varieties through selective breeding. Read a biography of Burbank and write a report describing some of his contributions to selective plant breeding.

Figure 4–3 *A mule (bottom) combines the best traits of a horse (top left) and a donkey (top right). What is this selective-breeding technique called?*

Cloning

A clone is an organism that is genetically identical to its parent. Seedless grapes and navel oranges are examples of clones. Use library reference materials to learn more about cloning. What are some methods of cloning? What kinds of organisms can be produced by cloning? Make a poster or bulletin board display to illustrate your findings.

In some ways, hybrid offspring may have traits that are better than those of either parent. The hybrid offspring may be stronger or healthier than its parents. Such offspring are said to have hybrid vigor. The word vigor means strength or health. Mules, for example, have more endurance than horses and are stronger than donkeys. One disadvantage of hybridization, however, is that the hybrid offspring is usually sterile, or unable to reproduce.

Inbreeding

Another selective-breeding technique is called **inbreeding.** Inbreeding is the opposite of hybridization. **Inbreeding involves crossing plants or animals that have the same or similar sets of genes, rather than different genes.** Inbred plants or animals have genes that are very similar to their

Figure 4–4 *As a result of inbreeding, all cheetahs are closely related and are susceptible to the same diseases. Why might this be hazardous to the cheetahs' survival?*

parents' genes. One purpose of inbreeding is to keep various breeds of animals, such as horses, pure. Purebred animals tend to keep and pass on their desirable traits. For example, a purebred racehorse that has won many races may be able to pass on its speed and strength to its offspring.

Unfortunately, inbreeding reduces an offspring's chances of inheriting new gene combinations. In other words, inbreeding produces organisms that are genetically similar. This similarity, or lack of genetic difference, in inbred plants and animals may cause the organisms to be susceptible to certain diseases or changing environmental conditions. For example, almost all cheetahs are genetically identical. If all cheetahs have nearly the same genes, they are all susceptible to the same diseases. As a result, wild cheetahs might eventually become extinct, or die off.

4–1 Section Review

1. What is selective breeding?
2. How is inbreeding different from hybridization?
3. What is one advantage of inbreeding? What is one disadvantage?

Critical Thinking—*Making Inferences*

4. Why do you think animals that are produced through inbreeding look so much alike?

ACTIVITY DOING

Examining Different Fruits

In this activity, you will examine a tangelo, which is a cross between a grapefruit and a tangerine. You will also compare the characteristics of a tangelo with those of a grapefruit and those of a tangerine.

1. Obtain a tangelo, a grapefruit, and a tangerine. Place each fruit on a paper towel.

2. Construct a data table to record the following traits for each fruit: size, color, seed size, juiciness, taste, odor.

3. Use a knife to cut each fruit in half. **CAUTION:** *Be careful when using a knife or any sharp instrument.*

4. Examine the tangelo, grapefruit, and tangerine for each trait listed in the data table. Record your observations.

What were the desirable traits in each fruit?

What were the undesirable traits in each?

Why are fruits such as the tangelo developed by plant breeders?

Guide for Reading

Focus on these questions as you read.

- *How does the process of genetic engineering use recombinant DNA?*
- *What are some products of genetic engineering?*

4–2 Genetic Engineering

At one time, most hybrid plants and animals were produced through selective-breeding techniques. In the not-too-distant future, **genetic engineering** may be the primary method of producing hybrids. **Genetic engineering is the process in which genes, or pieces of DNA, from one organism are transferred into another organism.** The production of the supermouse you read about at the beginning of this chapter is an example of genetic engineering.

Recombinant DNA

In one form of genetic engineering, parts of an organism's DNA are joined to the DNA of another organism. The new piece of combined DNA is called **recombinant DNA.** Pieces of recombinant DNA contain DNA from two different organisms. Usually, DNA is transferred from a complex organism (such as a human) into a simpler one (such as a bacterium or a yeast cell). Bacteria and yeast cells are used because they reproduce quickly. As the bacteria or yeast cells reproduce, copies of the recombinant DNA are passed on from one generation to the next. In each generation, the human DNA causes the bacteria or yeast cells to produce human protein.

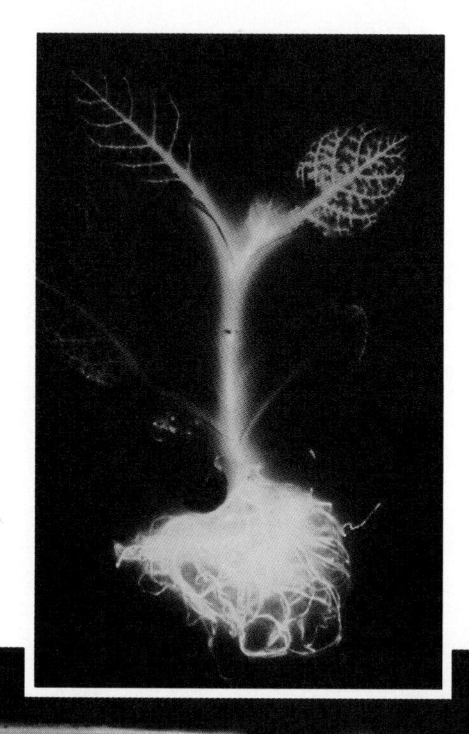

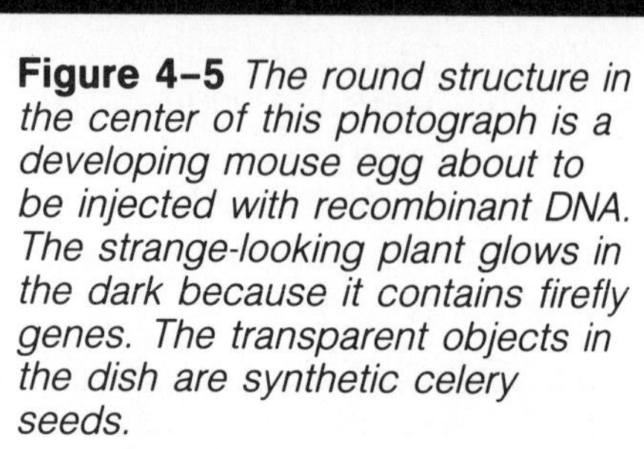

Figure 4–5 *The round structure in the center of this photograph is a developing mouse egg about to be injected with recombinant DNA. The strange-looking plant glows in the dark because it contains firefly genes. The transparent objects in the dish are synthetic celery seeds.*

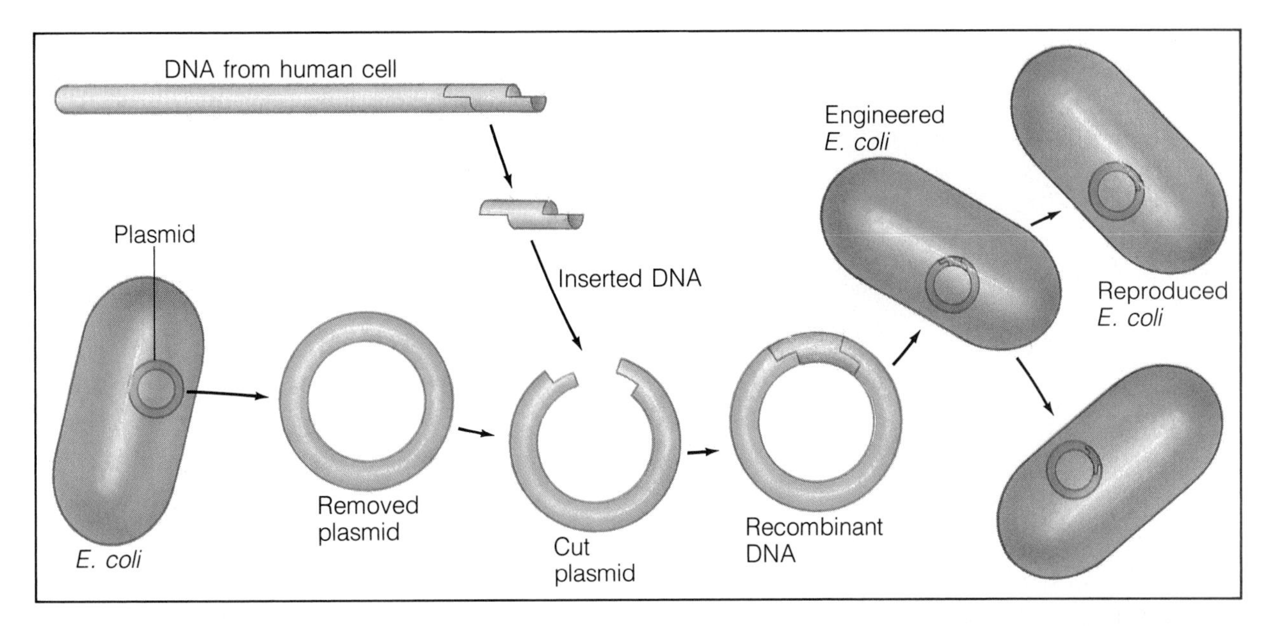

Figure 4–6 *To make recombinant DNA, a plasmid from a bacterium such as* E. coli *is snipped open. A short piece of DNA is then removed from a human cell. The human DNA is inserted into the cut plasmid. Then the plasmid is placed back into the bacterium. What happens next?*

Activity Bank

How Do Bacteria Grow?, p.112

Making Recombinant DNA

Scientists use special techniques to make recombinant DNA. Figure 4–6 illustrates this process using bacterial and human DNA. Refer to this diagram as you read the description that follows.

Some of the DNA in the bacterium *E. coli* is in the form of a ring called a **plasmid.** You might think of a plasmid as a circle of string. Using special techniques, scientists first remove a plasmid from a bacterium and cut it open. Then they remove a piece of DNA from a human cell. Think of this human DNA as a short piece of string. The scientists then "tie" this piece of human DNA to the cut ends of the bacterial DNA. The bacterial DNA again forms a closed ring. But the bacterial DNA ring now contains a human gene that directs the production of a human protein!

Finally, the scientists put the recombinant DNA back into the bacterial cell. What do you think happens next? The bacterial cell and all its offspring now produce the human protein coded for by the gene in the human DNA. In this way, large amounts of human protein can be produced outside the human body.

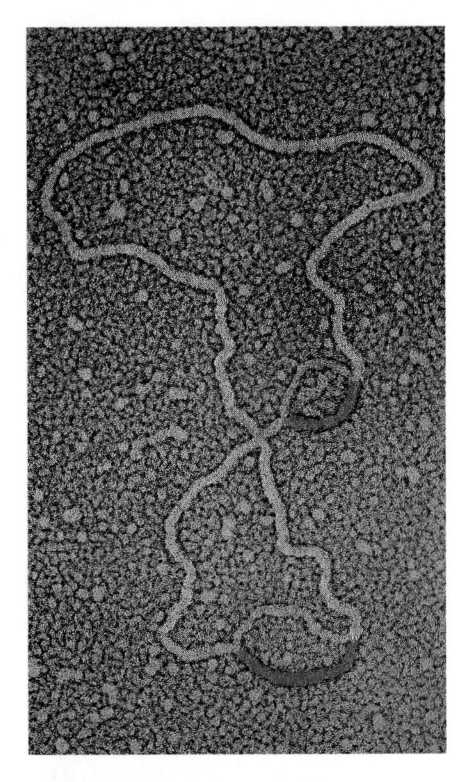

Figure 4–7 *Color has been added to this photograph of a bacterial plasmid to highlight two genes (red and blue sections of the plasmid).*

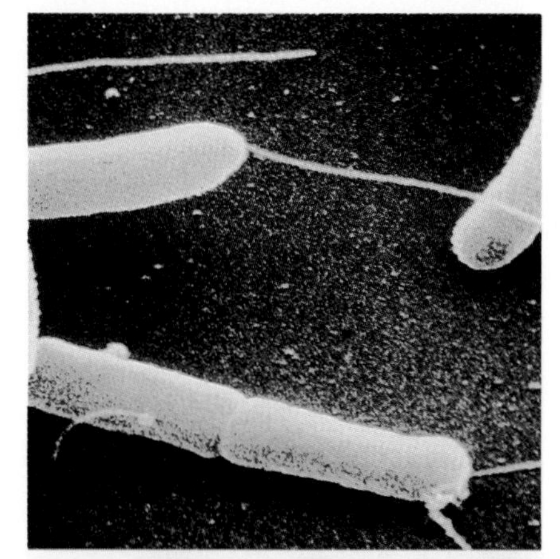

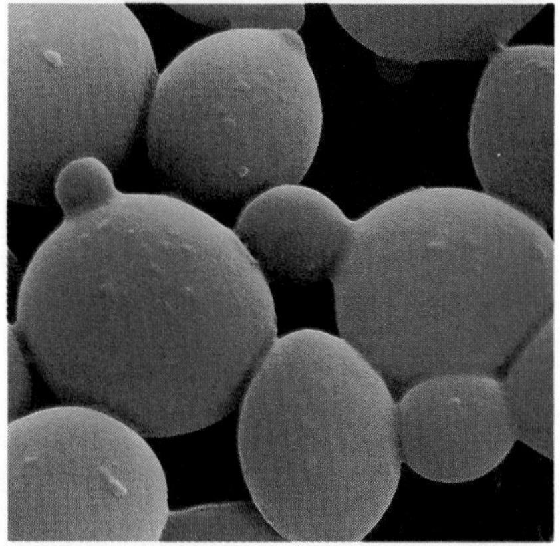

Figure 4–8 *Bacteria, such as* E. coli *(top), and yeast cells (bottom) are often used by scientists to make recombinant DNA. Why are bacteria and yeast useful for this purpose?*

Products of Genetic Engineering

Scientists use genetic engineering to turn certain bacteria and yeast cells into protein "factories." Grown in huge containers, billions of genetically engineered bacteria and yeast cells produce enormous quantities of proteins. These proteins have important uses in both medicine and agriculture. In medicine, the proteins are used to test for diseases such as AIDS, to treat human disorders such as diabetes, and to make vaccines that help fight diseases such as hepatitis B. In agriculture, genetic engineering is used to help make plants resistant to cold, drought, and disease.

MEDICINE One important product of genetic engineering is human insulin. Without this hormone, the level of sugar in the blood rises, causing a disorder called diabetes mellitus. Some people with diabetes must receive one or more injections of insulin daily. In the past, the insulin used to treat diabetes came from animals such as pigs and cattle. However, many people with diabetes were allergic to this animal insulin. In addition, supplies of animal insulin were limited and expensive. Today, human insulin is produced by genetically engineered bacteria. This insulin does not cause allergies in humans. Supplies are plentiful, and the insulin is inexpensive as a result.

Another protein made by bacteria through genetic engineering is human growth hormone. This hormone, which is normally produced by a gland near the brain, controls growth. A lack of human growth hormone prevents children from growing to their full height. Children whose bodies do not produce enough human growth hormone can be given injections of the hormone. These children often grow 6 to 8 centimeters more each year than they would without the injections of growth hormone. Until 1981, however, there was only a limited supply of human growth hormone available. Many children could not be treated. Then in 1982, bacteria were genetically engineered to produce human growth hormone. Now an almost unlimited supply is available.

Vaccines can also be produced through genetic engineering. When introduced into a person's body, a vaccine triggers the production of antibodies.

Return of the Dinosaurs

For a fictional account of how genetic engineering might help scientists recreate extinct dinosaurs, read *Jurassic Park* by Michael Crichton.

Figure 4–9 *Children normally grow at different rates. If a child's body does not produce sufficient amounts of human growth hormone, however, genetically engineered human growth hormone may be administered.*

Antibodies protect a person from disease. Vaccines are made from disease-causing viruses or bacteria. At one time, making the vaccine for hepatitis B (a serious liver disease) was expensive. Now scientists can remove a gene from the hepatitis B virus and insert the gene into a yeast cell. The yeast cell multiplies rapidly and makes large amounts of viral protein. The viral protein is then used to make hepatitis B vaccine. Hepatitis B vaccine is now less expensive to make than it was before genetic engineering.

Another product of genetic engineering is interferon. Interferon, a protein normally produced by human body cells, helps the body fight viruses. One form of interferon may even be helpful in fighting the virus that causes AIDS. Scientists are not really sure how interferon fights viruses. But they do know that when a virus enters a cell, it produces interferon. The interferon then leaves the infected cell and prevents the virus from infecting other cells. Interferon was once very expensive to make. But now, as a result of genetic engineering, supplies of interferon are less expensive and more plentiful.

AGRICULTURE A wide variety of viruses infect important crop plants, including wheat, corn, potatoes, tomatoes, and tobacco. For example, a virus called tobacco mosaic virus attacks and damages tobacco and tomato plants. Scientists have now found a way to protect these plants from the disease-causing virus. Using genetic engineering, scientists can insert genes from the tobacco mosaic virus into plant cells. Although scientists do not yet know why, the viral genes make the plant resistant to tobacco mosaic disease.

Figure 4–10 *You can see the effects of tobacco mosaic virus on the leaves of the tobacco plant. How might genetic engineering help to protect plants from being infected with the tobacco mosaic virus (inset)?*

Figure 4–11 *Are these strawberries ruined? Genetically engineered ice-minus bacteria might have saved them.*

Sub-Zero, p.113

Transgenic Crops

In the 1980s, scientists began growing genetically engineered, or transgenic, food plants in the laboratory or in small garden plots. Within a few years, transgenic tomatoes, corn, rice, and lettuce may be available in supermarkets. Some consumers are concerned that these transgenic foods may be dangerous to human health. Organize a class discussion of the pros and cons of genetically engineered food plants.

Another interesting use of genetic engineering in agriculture is the development of "ice-minus" bacteria. These genetically engineered bacteria help slow the formation of frost on plants. To understand how ice-minus bacteria work, you must first know something about the bacteria that normally live on plants. These normal bacteria are referred to as ice-plus bacteria. An ice-forming gene in ice-plus bacteria controls the production of a protein that triggers freezing. When the temperature drops to the freezing point of water (0°C), the water in plant cells freezes and turns to ice. The formation of ice in plant cells causes the cells to rupture and die. Many commercially important crop plants are ruined every year by frost damage.

When the ice-forming gene is removed from ice-plus bacteria, the protein that triggers freezing is not produced. Without the protein, ice still forms in plant cells, but at a lower temperature (−5°C). The bacteria are now called ice-minus bacteria. Scientists hope that some day soon crop plants such as strawberries and oranges will be protected from frost damage by using genetically engineered ice-minus bacteria.

Many environmentalists oppose the use of ice-minus bacteria. They are concerned that if the genetically engineered bacteria are released into the environment, they might turn out to be harmful. Further tests will be necessary before ice-minus bacteria can be used by farmers.

4–2 Section Review

1. What is genetic engineering?
2. What is recombinant DNA? Describe the process of making recombinant DNA.
3. Why are bacteria and yeast cells used to make large amounts of human proteins?
4. Describe two ways in which genetic engineering has been useful in medicine. In agriculture.

Connection—*You and Your World*

5. How are ice-minus bacteria similar to the antifreeze used in an automobile radiator?

CONNECTIONS

Frankenstein Fishes

"We're going to have Frankenstein fish!" exclaims one geneticist. "We're going to feed the world," says another. What are they talking about? Both scientists are discussing new, genetically engineered fishes that are larger and grow faster than normal fishes.

In 1985, scientists in China announced the transfer of a gene for human growth hormone into goldfish eggs. As the goldfish developed, some of them grew two to four times their normal size! Three years later, scientists in the United States transferred a growth gene from rainbow trout into another type of fish called carp. These carp grew 20 to 40 percent larger than usual. "Not only did they grow bigger and faster," reported the leader of the team of scientists, "but their offspring grew faster too."

Since these experiments were performed, fishes have been genetically altered in a variety of ways. In addition to growing bigger fishes, scientists are also experimenting with alterations that would make fishes resistant to diseases and pollutants in the *environment,* and also able to withstand very cold temperatures. Because fishes are an important source of food, these experiments could be important to commercial aquaculture, or fish farming.

Although it will be several years before genetically engineered fishes are available commercially, environmentalists are concerned that these fishes could cause problems in aquatic ecosystems. As of now, genetically engineered fishes are kept in aquaculture ponds or laboratory tanks. But what might happen if these fishes were released into the wild? No one knows. However, scientists, environmentalists, and government agencies agree on the need for safety guidelines and regulations to control the research and release of genetically engineered fishes.

Laboratory Investigation

Recombinant DNA

Problem

How can you make models to represent recombinant DNA?

Materials *(per group)*

construction paper (different colors)
tracing paper
drawing compass
tape
scissors

Procedure

1. Use a drawing compass to draw a circle 6 cm in diameter on a piece of construction paper.
2. Inside the large circle, draw a smaller circle 2.5 cm in diameter. You should now have a strip about 1.8 cm wide between the inner circle and the outer circle.
3. Using scissors, carefully cut around the outer circle and cut away the inner circle. **CAUTION:** *Be careful when using scissors or any sharp instrument.* You should now have a closed ring of construction paper.
4. Repeat steps 1 through 3 to make two more construction paper rings.
5. Trace each of the DNA segments shown here on a piece of tracing paper. Label each segment as shown.
6. Carefully cut out each DNA segment from the tracing paper. Using each DNA segment as a pattern, cut three DNA segments from construction paper. Use a different-colored piece of construction paper for each DNA segment.
7. Use the scissors and tape to make three models of recombinant DNA from the three closed rings and DNA segments. Refer to the sequence of steps described in Section 4–2 and Figure 4–6 on page 85 as a guide.

Observations

1. What do the rings of construction paper represent in your models?
2. What do the DNA segments represent?

Analysis and Conclusions

1. What human protein would each of your recombinant DNA molecules produce in a living organism?
2. **On Your Own** The technique of making recombinant DNA is sometimes called gene splicing. Do you think this is a good name? Why or why not? (*Hint:* Look up the word splice in a dictionary.)

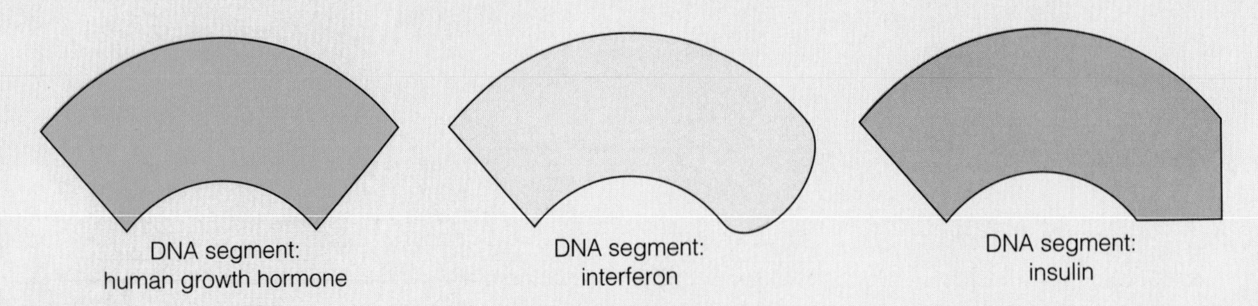

Summarizing Key Concepts

4–1 Plant and Animal Breeding

▲ Plant and animal breeders use selective breeding to produce offspring with desirable characteristics.

▲ Hybridization is a form of selective breeding in which two genetically different species are crossed.

▲ Hybrids are bred to have the best traits of both parents.

▲ Hybrids that are stronger or healthier than either parent are said to have hybrid vigor.

▲ Inbreeding is a form of selective breeding that involves crossing organisms with similar genes.

▲ Inbreeding produces organisms that are genetically similar.

▲ As a result of inbreeding, an offspring's chances of inheriting new genetic combinations is greatly reduced. This can make an entire inbred species susceptible to disease and could lead to extinction.

4–2 Genetic Engineering

▲ Through genetic engineering, genes, or pieces of DNA, are transferred from one organism to another organism.

▲ One form of genetic engineering involves the use of recombinant DNA, which contains pieces of DNA from two different organisms.

▲ Bacteria and yeast cells are commonly used in genetic engineering to produce human proteins.

▲ To make recombinant DNA, scientists remove a plasmid, or ring of DNA, from a bacterium and insert a piece of human DNA.

▲ As a result of genetic engineering, human proteins can be made outside the human body.

▲ Products of genetic engineering are used in medicine to produce hormones and vaccines and in agriculture to make plants resistant to disease and freezing.

Reviewing Key Terms

Define each term in a complete sentence.

4–1 Plant and Animal Breeding

selective breeding
hybridization
inbreeding

4–2 Genetic Engineering

genetic engineering
recombinant DNA
plasmid

Chapter Review

Content Review

Multiple Choice

Choose the letter of the answer that best completes each statement.

1. Crossing two genetically different plants or animals is called
a. inbreeding.
b. hybridization.
c. genetic engineering.
d. crossbreeding.

2. The word vigor in the term hybrid vigor means
a. offspring.
b. weakness.
c. strength.
d. trait.

3. Purebred plants and animals are produced through
a. inbreeding.
b. hybridization.
c. genetic engineering.
d. recombinant DNA.

4. Inserting genes from one organism into another is an example of
a. hybridization.
b. inbreeding.
c. crossbreeding.
d. genetic engineering.

5. To make recombinant DNA, human DNA is usually transferred into yeast cells or
a. mouse cells.
b. bacteria.
c. viruses.
d. plant cells.

6. Inbreeding produces organisms that are genetically
a. different.
b. similar.
c. identical.
d. opposite.

7. Genetic engineering can be used to produce
a. insulin.
b. human growth hormone.
c. interferon.
d. all of these.

8. The human protein needed to treat diabetes mellitus is
a. human growth hormone.
b. interferon.
c. insulin.
d. hemoglobin.

True or False

If the statement is true, write "true." If it is false, change the underlined word or words to make the statement true.

1. Hybridization is the process of crossing two genetically <u>similar</u> organisms.

2. A hybrid offspring combines the <u>worst</u> traits of both parents.

3. A method of selective breeding that is the opposite of hybridization is called <u>inbreeding</u>.

4. <u>Inbreeding</u> <u>increases</u> an offspring's chances of <u>inheriting</u> different gene combinations.

5. As a result of <u>hybridization</u>, some organisms might be <u>in danger of</u> becoming extinct.

Concept Mapping

Complete the following concept map for Section 4–1. Refer to pages E6–E7 to construct a concept map for the entire chapter.

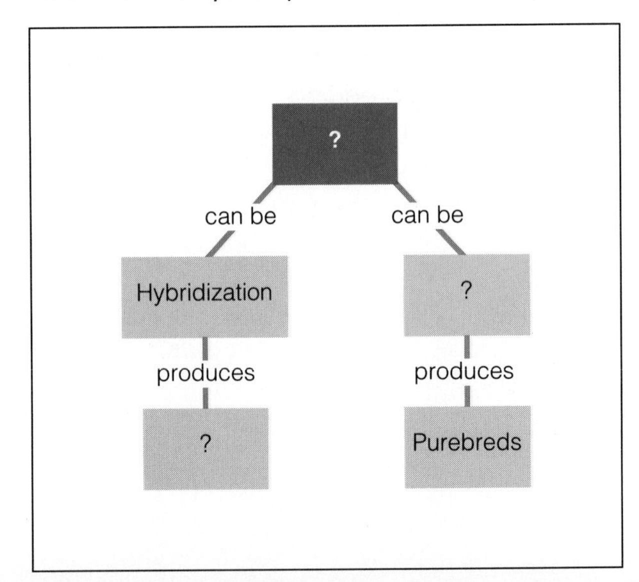

Concept Mastery

Discuss each of the following in a brief paragraph.

1. How do breeders produce organisms with desired characteristics?
2. What is genetic engineering? How has genetic engineering affected modern life?
3. Describe the steps involved in the production of recombinant DNA.
4. Explain how bacteria and yeast cells can be used as "factories" to make human proteins.
5. What is the advantage of developing ice-minus bacteria? Are there any disadvantages? Explain.
6. What is one disadvantage of inbreeding? Give an example.
7. Describe three applications of genetic engineering in medicine.
8. Describe how selective plant breeding was first used in agriculture.

Critical Thinking and Problem Solving

Use the skills you have developed in this chapter to answer each of the following.

1. **Sequencing events** The diagram below shows the steps involved in the process of making recombinant DNA. However, the steps are out of order. Place the steps in the proper sequence.

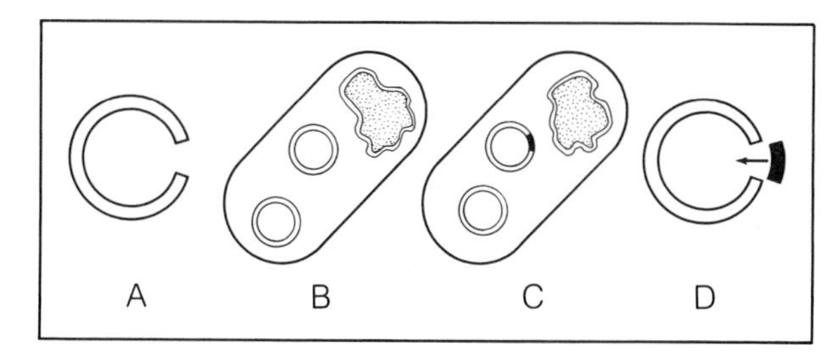

2. **Relating concepts** Do you think you would be able to predict the traits of the puppies resulting from a cross between two mixed-breed dogs? Why or why not?
3. **Applying concepts** Suppose you were asked to develop a wheat plant that can grow in a cold, dry environment and that is resistant to disease. How would you go about developing such a plant? What characteristics would you want your wheat plant to have? Why would these characteristics be important?
4. **Making inferences** Scientists working with ice-minus bacteria discovered that the ice-forming proteins produced by ice-plus bacteria could be used to help make snow at ski resorts. Ski resorts now use these proteins to make snow faster and in warmer weather. Do you think that releasing ice-forming proteins into the environment caused the same concerns as releasing ice-minus bacteria? Why or why not?
5. **Using the writing process** Many people are concerned about the introduction of new, genetically engineered organisms into the environment. These people think that genetically engineered organisms may be harmful to the environment or to other living things. What safety guidelines would you recommend concerning the possible development of genetically engineered plants and animals? Write an essay describing your guidelines and giving your reasons for including each guideline.

Barbara McClintock: She Discovered "JUMPING" GENES

The news headline for October 10, 1983, read "Biologist Wins Nobel in Medicine." Eighty-one-year-old Barbara McClintock had just won the world's greatest scientific award. She received the award for her discovery that genes can move from one spot to another on a chromosome—or even from one chromosome to another. Many people felt that the award was long overdue. For Barbara McClintock had made the discovery thirty years earlier.

It had taken the scientific world that long to realize the importance of McClintock's research. In 1951, when Dr. McClintock first reported her discovery, she was met with silence. Her fellow scientists either did not understand or would not believe that genes do not always remain in a fixed spot on a chromosome.

When McClintock began her research, scientists did not have the knowledge or equipment to unravel the chemical makeup of genes. To determine how genes work in plants, McClintock decided to examine changes in the outward appearance of plants. McClintock used the maize plant to study the color variations in kernels. (Maize is another name for corn.)

Dr. McClintock's early research during the 1920s and 1930s proved that genes determine a maize plant's characteristics, such as color. In further studies during 1944–45, she observed a pattern of color on some corn kernels that was unlike anything she had seen before. Dr. McClintock wondered how this could be explained.

To find the explanation, McClintock studied the color patterns of kernels in many generations of corn. Using a microscope, she studied certain genes on the chromosome of each plant. These genes controlled changes in the color of kernels. She tried to match the color of the kernels with the position of these genes on their chromosomes.

▼ **A very happy Dr. McClintock holds a sample of the corn that helped her win a Nobel prize.**

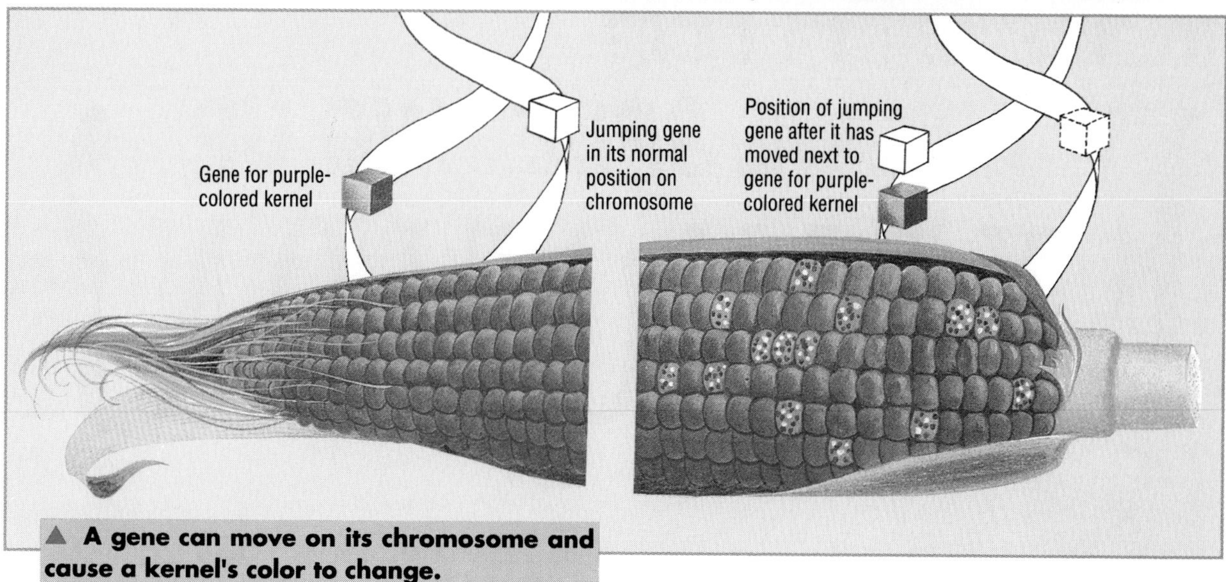

▲ **A gene can move on its chromosome and cause a kernel's color to change.**

Some genes, she found, moved! And when these genes moved, they caused the color patterns of the corn kernels to change. How did the jumping genes do this?

When a jumping gene moved, it landed next to genes that controlled kernel color. The jumping gene then affected the action of these coloring genes. For example, let's say the nearby genes caused a kernel to be purple. The jumping gene interfered with the action of the "purple kernel" genes. This resulted in changes in the color of the corn kernel. Instead of being completely purple, the kernel was now speckled purple, pink, and white.

This discovery of jumping genes did not fit in with older scientific ideas about genes. Scientists believed that genes stayed in one place on the strands of chromosomes. It was as if the genes were beads on a necklace. But McClintock discovered that some genes, at least, could move on the chromosomes. When the genes did this, mutations, or changes, were caused in the organisms.

Despite the fact that her findings at the time were ignored, McClintock continued her work. Her confidence was unshakable. As she said, "If you know you are on the right track, if you have this inner knowledge, then nobody can turn you off . . . regardless of what they say."

Finally, during the late 1960s, researchers began to find jumping genes in other organisms, such as bacteria. Passed on to new generations of bacteria, some jumping genes could give disease-causing bacteria the ability to resist antibiotic medicines.

Jumping genes in the one-celled parasite that causes African sleeping sickness help the parasite to overcome a person's natural resistance to the disease. More recent studies seem to show that movable genes may also create changes in normal cells that turn them into cancer cells.

Such findings prompted scientists to recognize at last the importance of McClintock's earlier discovery. As Dr. James Watson, one of the discoverers of the structure of DNA, said, "It's really that science has caught up with Barbara."

By the 1970s, Barbara McClintock had become something of a scientific hero. Yet she continued her quiet life at the research center at Cold Spring Harbor on Long Island, New York. For more than 50 years, she worked 12 hours a day, 6 days a week in her laboratory. On September 2, 1992, at the age of 90, Barbara McClintock died.

Dr. Watson, director of the Cold Spring Harbor Laboratory, called McClintock one of the three most important figures in the history of genetics, one of "the three M's." The others are Gregor Mendel and Thomas Hunt Morgan. Although she was only a little more than five feet tall, Dr. McClintock will be remembered as a giant in genetics.

ARE WE *SPEEDING* UP EXTINCTION?

About 70 million years ago, the last dinosaur to roam the Earth took its final breath and died. What kind of dinosaur was this lone survivor? Perhaps it was the mighty *Tyrannosaurus*, a fierce flesh eater. Or maybe it was the three-horned *Triceratops*, a powerful plant eater. Possibly it was one of the many smaller types of dinosaurs that skittered over the landscape on long hind legs. No one knows what kind of dinosaur was the last to look upon the Earth. Its identity will forever remain a mystery. But one fact is clear: With the death of the last one, dinosaurs became extinct and would not be seen again.

Conservationists disagree on whether many species of plants and animals are becoming extinct, or dying out, as part of the natural order of things. Some feel that people are making the environment unfit for certain plants and animals.

You may hear the word extinction often these days. Many types of plants and animals have vanished in recent times. With the loss of the rain forests, many more are on the brink of disappearing. Since the year 1600, about 1000 kinds of mammals, birds, and other vertebrates (animals with backbones) have become extinct. Extinction has always been a fact of life. But has this process been speeded up by the interference of people? Are living things dying off at an unnaturally fast rate?

A report from the Florida Conservation Foundation states that "extinction is the ultimate fate of all species." However, the report goes on to say that "the modern rate of extinction is not natural." If this is true, what has caused the rapid disappearance of so many organisms? And why are thousands more on the endangered species list?

One possible answer has been suggested by S. Dillon Ripley, former secretary of the Smithsonian Institution. Ripley, like many conservationists, feels that certain organisms are in danger because of "man and his intrusion into their fragile environments."

Some scientists disagree with this position on the role of people in the extinction process. Among them is Dr. John J. McKetta, a chemical engineering professor at the University of Texas, who wrote that "it is possible that . . . man may hasten the disappearance of certain species. However, the evidence indicates that he has very little to do with it."

Few conservationists would agree with McKetta's view. But many admit that some species are so primitive that they are easy targets for extinction. This may be the opinion of James Fischer, Noel Simon, and Jack Vincent, three noted conservationists. They have written that "in any period, including the present, there are doomed species: naturally doomed species bound to disappear."

The African black rhinoceros has been brought to near extinction by hunters. Will its fate be that of the dinosaur, who mysteriously disappeared millions of years ago?

▲ **Animals that can adapt to various conditions are more likely to survive in a changing environment. Raccoons, for example, are country animals that easily adapt to city life.**

It cannot be argued that the types of organisms that survive are those that adapt best to the changing conditions of the environment. The surroundings in which an organism lives change all the time. Some changes are big; others small. Some changes happen quickly; others slowly. Many of these changes are natural. Mountains rise up, and old mountains crumble. Ponds dry up. Floods cover forest areas. New species appear and compete with the old for food and space.

But conservationists are concerned that when people tamper with an environment, many plants and animals cannot adjust. Although they may be able to adapt to natural changes, some organisms cannot keep pace with changes caused by humans.

The African black rhinoceros is an example of an animal threatened with extinction by humans. Huge and powerful, the rhino has few natural enemies. It is adapted to living on vast plains, eating the tough plants that grow there. But the rhino is a slow-witted creature and is easy prey for human hunters. Despite the laws that are supposed to protect rhinos, hunters kill the animals for their horns. Rhino horns bring high prices in parts of the world where they are used in folk medicine. As a result, the rhinoceros is on the brink of extinction.

Similarly, the grizzly bear is a fierce animal with few enemies in the wild. For hundreds of years, grizzlies roamed the vast North American wilderness. But as farms, ranches, and towns began to replace wilderness, grizzlies began to decrease in numbers. Those that remained live much closer to people. And grizzlies and people do not mix well. Grizzlies seldom attack people or livestock. But when they do, they are usually hunted down and destroyed. Grizzly bears could be headed for extinction.

Both rhinos and grizzlies are what scientists call highly specialized creatures. This means that they have adapted to very special conditions. When people disturb these conditions—by shrinking the wilderness, for example—these animals find it difficult to exist.

Some animals, on the other hand, are not specialized. The are suited to living under many different kinds of conditions. The raccoon is one of these animals. Raccoons are adapted to the wild. But they also manage to survive well in cities. Instead of eating natural food, they feed on garbage. If they cannot find dens in hollow trees and logs, they make their homes in attics, abandoned buildings, and garages. Raccoons in Cincinnati, Ohio, even learned to use an underpass to cross a busy highway. While some animals have dwindled in number, raccoons are as abundant as ever.

Some scientists think that eventually only very adaptable species such as the raccoon will be able to survive in a world changed by people. Will new species develop that are more adaptable to the modern world? Have the rhino and the grizzly outlived their time? Conservationists suggest that if the changes caused by people drive wild creatures out of existence, the environment itself is not a healthy place. In the long run, a sick environment endangers the human species. Only time will tell whether or not people are endangered by an environment of their own making.

MAPPING THE HUMAN GENOME

The disembodied nose of an imaginary world leader plays an important role in a 1970s comic film. The movie's villains want to use the organ to recreate their dead leader, and the hero must steal—and eventually squash—the nose in order to stop them. The movie is more concerned with spoof than science, but there is some biological fact behind the humorous fiction. Indeed, every cell in a person's body—whether it is a nose cell, a liver cell, or a tongue cell—contains a complete set of instructions for making that person. And although scientists may not be interested in reviving dead dictators, many are deeply committed to understanding that set of genetic instructions.

For several years, scientists around the world have been working to decode the messages hidden in the human genome—the complete set of 46 chromosomes that determine who we are and how we develop. The project involves two basic, demanding tasks: mapping and sequencing. Scientists want a complete map of the genome, a map that will show the location of the approximately 100,000 genes on the chromosome strands. But they want to know the order of the DNA bases that make up those genes as well. Genome sequencing is the process by which scientists list, in careful order, the 3 billion to 6 billion base pairs that compose chromosomal DNA. Such a list would fill a book 1 million pages long!

Mapping and sequencing—not to mention organizing and cataloging all the data—will obviously take a long time and require a lot of money. Genome researchers will need support and funding from the scientific community as well as from the government. Many scientists, including DNA co-discoverer James Watson, are in favor of the genome project and defend the time and money ($200 million a year over a 15-year period) required for it. To these scientists, a complete genome map is the "Holy Grail" of biology. They say that the potential benefits of deciphering the genome are worth the cost of all the research. They expect that a better understanding of the genetic code will allow them to detect genetic flaws and develop more effective strategies for treating them. One example of such an advance is the recent discovery of the gene that goes wrong when a colon cell becomes cancerous. With this new finding, scientists hope to be able to detect colon cancer at its earliest stage—or even to identify people who have a predisposition for the disease. Researchers also anticipate using genetic engineering techniques and gene therapy to combat other killers, such as heart disease and AIDS. Scientists add that a complete knowledge of the genome may be the key to unlocking such biological mysteries as human behavior and evolution.

Opponents of the human genome project, however, insist that the drawbacks of genetic engineering outweigh its advantages. High on their list of concerns are ethical issues associated with probing the human genome. Some researchers and others fear that gene mapping and sequencing will lead to technologies that assert human control over natural processes. They worry that what begins as gene therapy to treat diseases will become gene enhancement to create "improved" individuals—people who are smarter, taller, fairer. These opponents argue that genome exploration is "bad science"—monotonous work with little value, which will drain funds and energy that might be better spent elsewhere. And they want at least to slow down the pace of gene mapping and sequencing.

▼ **Unlocking the mystery of the human genome might ensure that all babies grow up to fulfill their genetic potential.**

▶ Human body cells contain 23 pairs of chromosomes. Within each chromosome is a single molecule of DNA. The DNA molecule is a twisted ladderlike structure whose rungs are made of the nitrogen bases adenine (A), guanine (G), cytosine (C), and thymine (T).

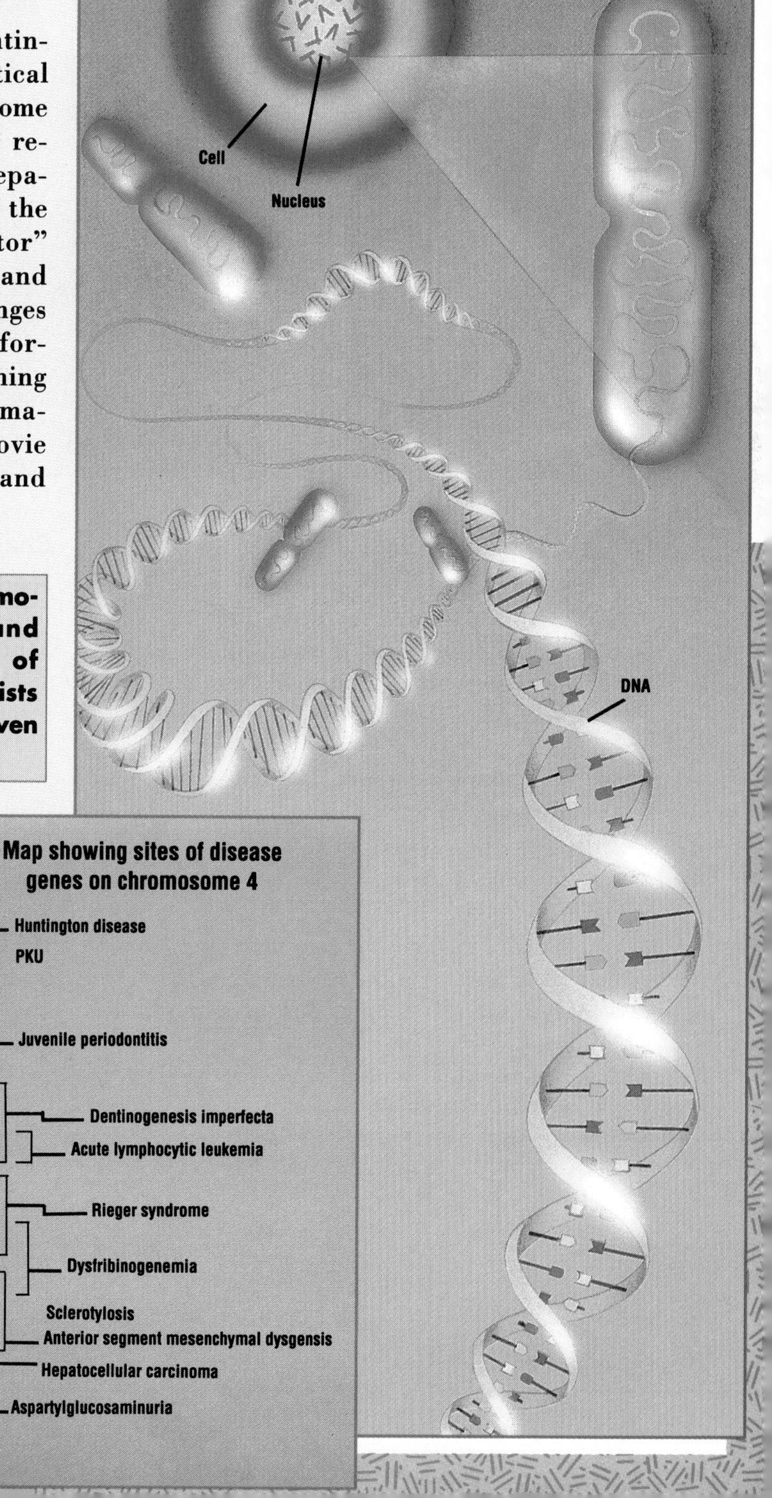

While the human genome project continues to stir ethical, economic, and political debate, however, scientists probe the genome and invent new techniques to aid their research. The discovery of methods for separating and arranging pieces of DNA and the development of an automatic "sequenator" make mapping and sequencing easier and more accurate. One of the biggest challenges of the project is managing the new information. Computer specialists are designing data banks to gather and organize information as it comes in. Who knows what movie makers will do with these issues, ideas, and inventions in the future?

▼ Some of the genes lined up on chromosomes may cause human diseases and disorders. By mapping the sequence of genes on different chromosomes, scientists hope some day to be able to treat—or even cure—some of these diseases.

Map showing sites of disease genes on chromosome 3

von Hippel-Lindau syndrome
Thyroid hormone resistance
Small cell cancer of lung
GM1-gangliosidosis
Renal-cell carcinoma
Protein S deficiency
Oroticaciduria
Propionicacidemia
Atransferrinemia
Postanesthetic apnea
Sucrose intolerance

Map showing sites of disease genes on chromosome 4

Huntington disease
PKU
Juvenile periodontitis
Dentinogenesis imperfecta
Acute lymphocytic leukemia
Rieger syndrome
Dysfribinogenemia
Sclerotylosis
Anterior segment mesenchymal dysgensis
Hepatocellular carcinoma
Aspartylglucosaminuria

For Further Reading

If you have been intrigued by the concepts examined in this textbook, you may also be interested in the ways fellow thinkers—novelists, poets, essayists, as well as scientists—have imaginatively explored the same ideas.

Chapter 1: What Is Genetics?

Chetwin, Grace. *Gom on Windy Mountain.* New York: Lothrop, Lee & Shepard.

Greenfield, Eloise, and Lessie Jones Little. *Childtimes.* New York: Thomas Y. Crowell.

Mayne, William. *Gideon Ahoy!* New York: Delacorte Press.

Wismer, Donald. *Starluck.* Garden City, NY: Doubleday & Co.

Chapter 2: How Chromosomes Work

Arkin, Alan. *The Lemming Condition.* New York: Harper & Row.

Lowry, Lois. *Rabble Starkey.* Boston: Houghton-Mifflin.

Paterson, Katherine. *Jacob Have I Loved.* New York: Thomas Y. Crowell.

Slepian, Jan. *The Alfred Summer.* New York: Macmillan Co.

Chapter 3: Human Genetics

Hamilton, Virginia. *Arilla Sundown.* New York: Greenwillow.

Skurzynski, Gloria. *Manwolf.* New York: Houghton Mifflin-Clarion Books.

Sleator, William. *Singularity.* New York: E. P. Dutton.

Chapter 4: Applied Genetics

Ames, Mildred. *Anna to the Infinite Power.* New York: Scribner.

Babbitt, Natalie. *Tuck Everlasting.* New York: Farrar, Straus and Giroux.

Hilton, James. *Lost Horizon.* New York: Morrow.

Activity Bank

Welcome to the Activity Bank! This is an exciting and enjoyable part of your science textbook. By using the Activity Bank you will have the chance to make a variety of interesting and different observations about science. The best thing about the Activity Bank is that you and your classmates will become the detectives, and as with any investigation you will have to sort through information to find the truth. There will be many twists and turns along the way, some surprises and disappointments too. So always remember to keep an open mind, ask lots of questions, and have fun learning about science.

TULIPS ARE BETTER THAN ONE

Gregor Mendel was able to cross pea plants with different traits through the process of cross-pollination. In cross-pollination, pollen from the stamen of one flower is transferred to the pistil of another flower. In addition to stamens and pistils, what are the other parts of a flower? This activity will help you to find out. You will need a tulip flower and a hand lens. Now follow these steps.

1. Compare your tulip flower with the drawing of a typical flower shown here.
2. Identify the sepals on the underside of the flower. Record their number, color, and shape. Break off a sepal and examine it with a hand lens.
3. Examine the flower's petals, the colored parts just above the sepals. Record their number, color, and shape. Remove a petal and examine it with a hand lens. Does your flower petal have a fragrance?
4. Identify the stamens. Record their number, color, and shape. Remove a stamen and examine it with a hand lens. Identify the filament, which is the stalklike structure, and the anther, the structure at the tip of the filament. Where is the pollen located? Shake some pollen onto a sheet of paper and examine the pollen with a hand lens. Draw what you see.
5. Locate the pistil at the center of the flower. Refer to the drawing of a typical flower to help you identify the stigma, style, and ovary. Where do the seeds develop?

Do It Yourself

Did you know that some plants have both male and female flowers? The male flowers contain only stamens and the female flowers contain only pistils. Go to the library and find out the names of several plants that have male and female flowers. If possible, obtain a specimen of such a flower and show it to the class.

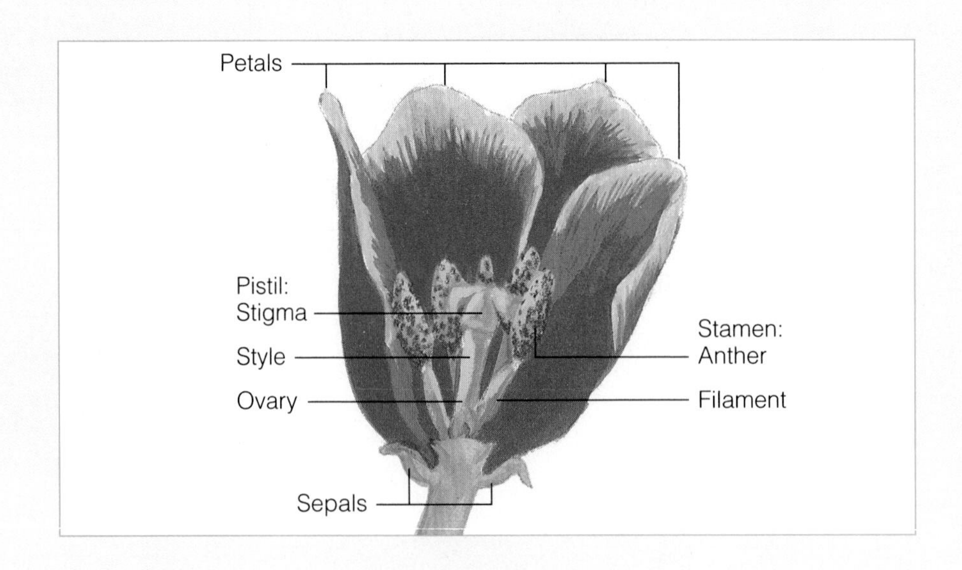

FLIP OUT!

Like Gregor Mendel, geneticists use the laws of probability to predict the results of genetic crosses. Do you ever use probability in your daily life?

Imagine that your school soccer team is about to play its first home game. The stands are filled to capacity. As team captain, you are about to take part in the coin toss. Nervously, you watch as the referee flips the coin into the air. "Heads!" you shout. Did you win the toss?

Probability is the likelihood that a coin will come up heads or tails on any one toss. What is the likelihood that the coin would land heads up and you would win the toss? You can perform a simple activity to find out. All you need is a coin.

First answer these questions. What are the possible ways the coin could have landed after it was tossed? What are the chances that the coin would have landed heads up? Tails up?

Suppose you were to toss a coin 20 times. How many times do you predict the coin would land heads up? Tails up? What percentage of the time do you predict the coin would land heads up? Tails up?

Now test your predictions. Flip the coin 20 times. Record the number of times it lands heads up and the number of times it lands tails up. Find the percentages by dividing the number of times the coin landed heads up or tails up by 20 and then multiplying by 100. What percentage of the time did the coin land heads up? Tails up? Were your predictions correct? Explain.

Combine the results of the coin toss for the entire class. Are the combined results closer to your predictions? Explain.

A MODEL OF MEIOSIS

Meiosis is the process in which an organism's sex cells are produced. The process of meiosis ensures that each sex cell has half the normal number of chromosomes for that particular organism. In this activity you will construct a model of meiosis for an organism with four chromosomes in its body cells.

Materials

3 sheets of paper
drawing compass
4 red pipe cleaners
4 white pipe cleaners

Procedure

1. On a sheet of paper, draw a circle about 15 cm in diameter. This circle represents a parent cell about to undergo meiosis.
2. Arrange four pipe cleaners, two white and two red, randomly inside the circle. The pipe cleaners represent two pairs of chromosomes.

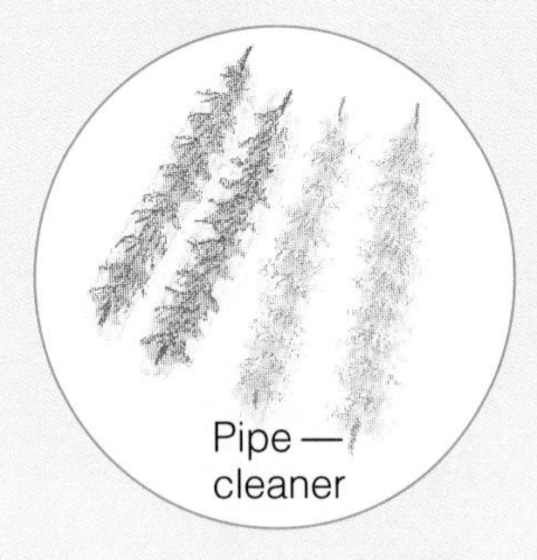

3. Place another white pipe cleaner next to each of the white "chromosomes." Place another red pipe cleaner next to each red "chromosome." These pipe cleaners represent the doubled chromosomes in the first step of meiosis.

4. Draw two circles side by side on a second sheet of paper. These circles represent the two new cells produced during the first cell division of meiosis. Divide the doubled "chromosomes" equally between the two new cells. How many white "chromosomes" are present in each new cell? How many red?

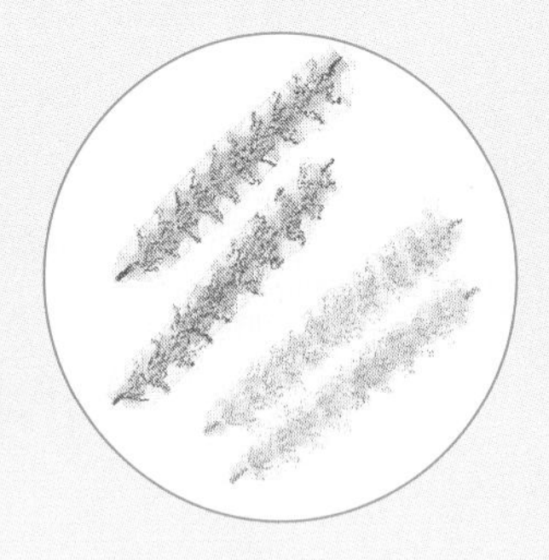

5. Draw four circles side by side on a third sheet of paper. These circles represent the sex cells that are the product of the second cell division of meiosis. Divide the pipe cleaners equally to represent the "chromosomes" present in the sex cells. How many white "chromosomes" are there in each sex cell? How many red?

Think for Yourself

1. Why is it important that each sex cell have only half the normal number of chromosomes found in body cells?
2. The process of meiosis is sometimes called reduction division. Do you think this term accurately describes the process? Why or why not?

Going Further

Make a model of meiosis for an organism with six chromosomes. How many more pipe cleaners will you need? Will you need to use a third color in addition to white and red? Why or why not?

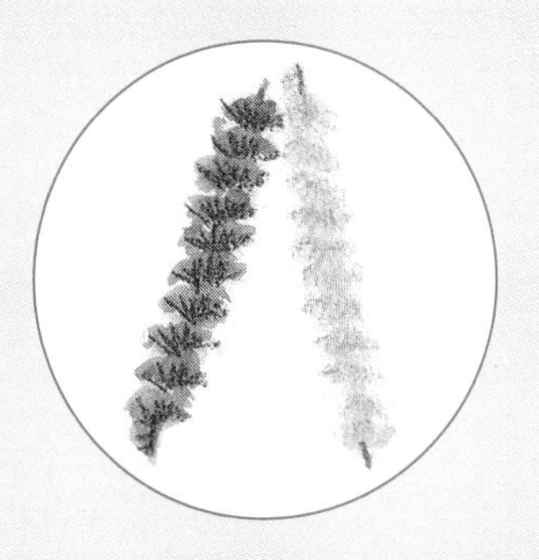
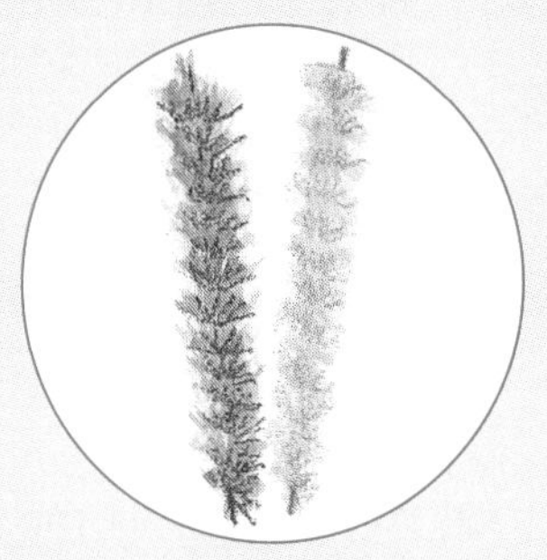
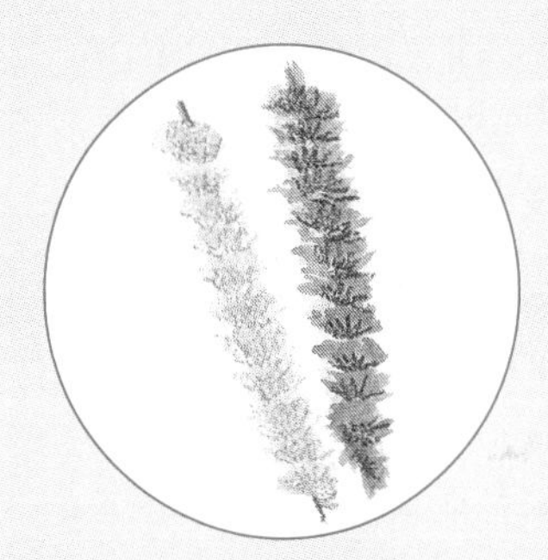
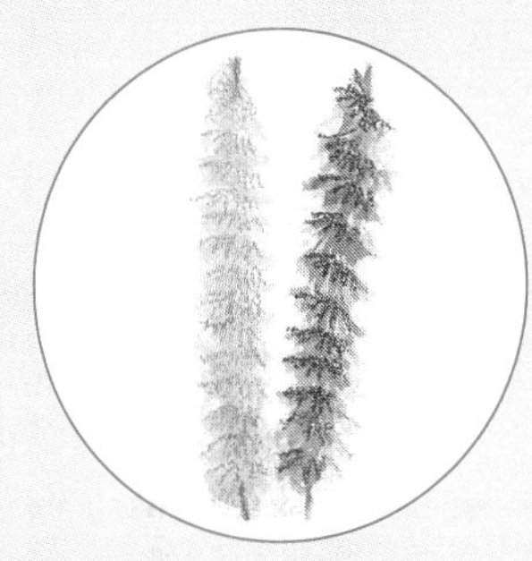

STALKING THE WILD FRUIT FLY

Ever since the days of Thomas Hunt Morgan, fruit flies have been used for genetic experiments. These tiny flies are commonly seen on and around displays of fresh (and overripe) fruits and vegetables. In this activity you will collect fruit flies and observe the stages in their life cycle.

Materials

paper towel	ripe banana
2 glass jars, 1 with cover	cotton
	hand lens

Procedure

1. Place a small piece of paper towel on the bottom of each jar.
2. Put a piece of ripe banana on the paper towel in each jar. Cover one jar tightly. Leave the second jar open.
3. Place both jars outdoors where they will not be disturbed for 24 hours.
4. After 24 hours observe the jars. Were fruit flies present in either jar? If so, which one?
5. Plug the mouth of the open jar with cotton. Use the hand lens to observe the adult fruit flies in the plugged jar. Can you see a difference between male and female fruit flies? Describe any differences you see.
6. Look for eggs that may have been deposited on the bottom of the jar. Describe the color and size of the eggs.
7. A few days after the eggs appear, look for tiny wiggly "worms" to emerge. These are fruit fly larvae, the second stage in their life cycle.
8. After a few days, watch for the larvae to crawl onto the paper and enter the pupa stage. The adult fruit flies will emerge from the pupae. How long did it take for the fruit flies to develop from eggs to adults?

Analysis and Conclusions

1. Describe the life cycle of fruit flies from egg to adult.
2. According to the old theory of spontaneous generation, fruit flies developed from rotting fruit. How does this activity disprove the theory of spontaneous generation?

WHERE DO PROTEINS COME FROM?

Your physical traits—from the shape of your ears to the color of your eyes—are determined by the chromosomes you received from your parents. The main function of chromosomes is to control the production of proteins, such as the enzymes that control eye color. Proteins are made in the cells of your body. Proteins are also present in the food you eat. Which foods contain proteins? You can perform a simple experiment to find out.

Materials

5 test tubes
5 rubber stoppers
glass-marking pencil
test-tube rack
graduated cylinder
Biuret solution
egg white
cottage cheese
bacon fat
canned tuna
milk

Procedure

1. Place a small sample of each of the foods listed into a separate test tube. Use the glass-marking pencil to label each test tube.

2. Add 5 mL of Biuret solution to each test tube. **CAUTION:** *Be careful not to spill Biuret solution on your skin or clothing. If a spill occurs, rinse with plenty of water.*

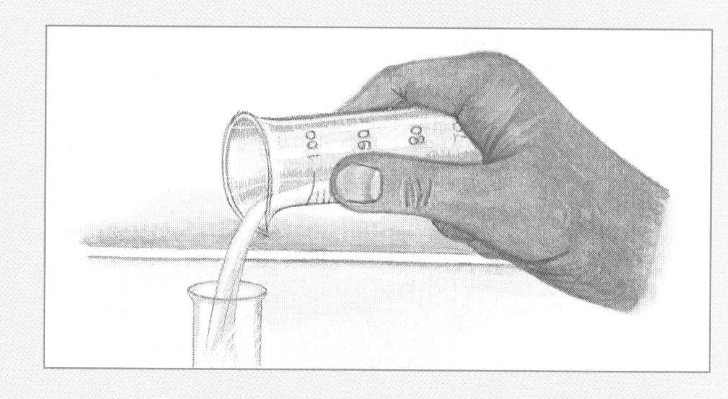

3. Seal each test tube with a rubber stopper. Shake each test tube and observe what happens. A pink or purple color indicates the presence of protein. If you do not see any color change, no protein is present. Record your observations in a data table similar to the one shown here. Which foods contain protein? Which do not?

Observations

DATA TABLE

Food	Protein Present (Yes/No)
Egg white	
Cottage cheese	
Bacon fat	
Tuna	
Milk	

Going Further

Do you think proteins are present in fruits, vegetables, and cereals? With your teacher's permission, design and perform an experiment to find out.

HOW CAN YOU GROW A PLANT FROM A CUTTING?

Once plant breeders have developed plants with desirable traits, it is important that they be able to produce more of the plants. One way they do this is to take a cutting from the original plant and let the cutting grow into a new plant. The new plant will be identical to the parent plant. Like plant breeders, you too can grow a plant from a cutting.

Materials

houseplant	knife
small pot	pencil
peat moss	plastic bag
coarse sand	rubber band

Procedure

1. Mix equal amounts of peat moss and coarse sand. Fill a small pot with this mixture to just below the rim of the pot.
2. Using a sharp knife, cut off the top 7 to 10 cm of the stem or side shoot of a houseplant. **CAUTION:** *Be careful when using a knife or other sharp instrument.*

3. Pull off the lower leaves and make a clean cut across the stem just below a leaf node.
4. With a pencil, make a hole about 3 cm deep in the potting mixture. Make the hole near the edge of the pot.

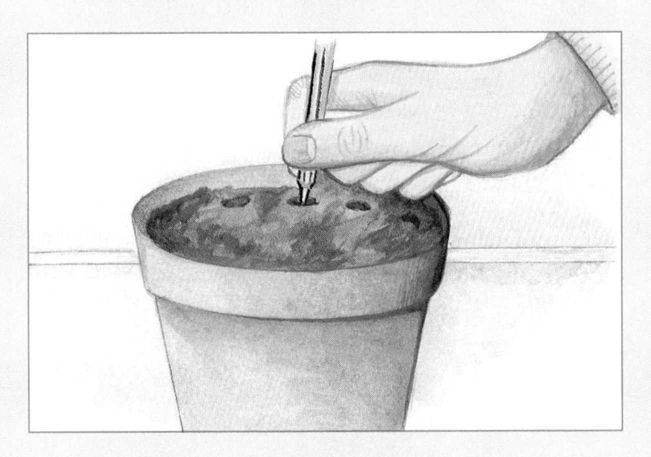

5. Insert the cutting so that the stem is supported by the edge of the pot. Gently firm the mixture around the cutting. **Note:** *You may want to make holes for several cuttings in the same pot.*

6. Water the cuttings thoroughly and let the pot drain.

7. Cover the pot with a plastic bag. Hold the bag in place with a rubber band.

8. Put the pot in a warm, shaded spot. Keep the mixture moist.
9. After three to four weeks, you should see new growth at the tips of your cuttings. Remove the plastic bag and carefully tilt the pot to remove the cuttings. Separate the cuttings and plant them in individual pots. Water your new plants and watch them grow!

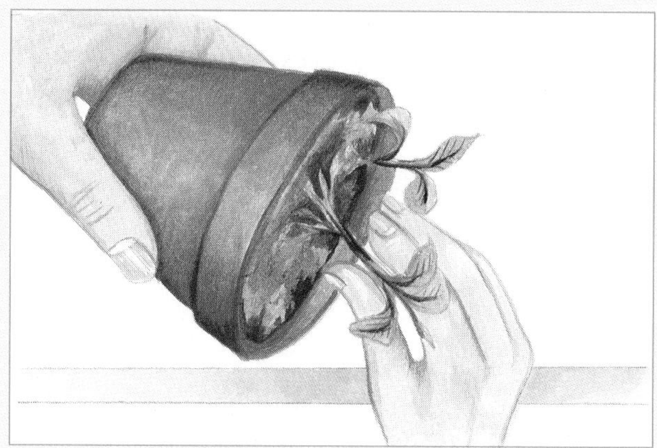

Do It Yourself

Growing a new plant from a cutting, as you did in this activity, is called vegetative propagation. Using reference materials, look up the meaning of this term. What are some other methods of vegetative propagation? Report on your findings to the class.

HOW DO BACTERIA GROW?

Bacteria are useful in genetic engineering because they reproduce quickly. Bacteria are single-celled organisms. They reproduce by splitting in two. This method of reproduction is called binary fission. ("Binary" means two and "fission" means to split.) In this activity you will observe how bacteria can be grown on agar, which is a substance made from seaweed.

Materials

2 sterile petri dishes with agar
glass-marking pencil
cotton swab
tape
hand lens

Procedure

1. Obtain two sterile petri dishes with agar. Why is it important that the petri dishes and agar be sterilized?
2. With a glass-marking pencil, label one petri dish A and the other dish B.
3. Draw a cotton swab across your desk, the back of your hand, a windowsill, or any other spot where you think bacteria might be present.
4. Raise one side of the lid of petri dish A. Move the cotton swab in a zigzag motion across the surface of the agar. Immediately close the lid of the petri dish. Why is it important that you raise only one side of the lid and then close it immediately?
5. Do not open petri dish B. Tape both petri dishes closed. What is the function of petri dish B?
6. Place both petri dishes in a warm, dark place where they will not be disturbed.
7. Observe each dish with a hand lens every day for four days. **CAUTION:** *Do not open the petri dishes.* Record your observations in a data table similar to the one shown here.
8. After four days, draw what you see in petri dish A and in petri dish B. In which petri dish did you see more bacteria growing?

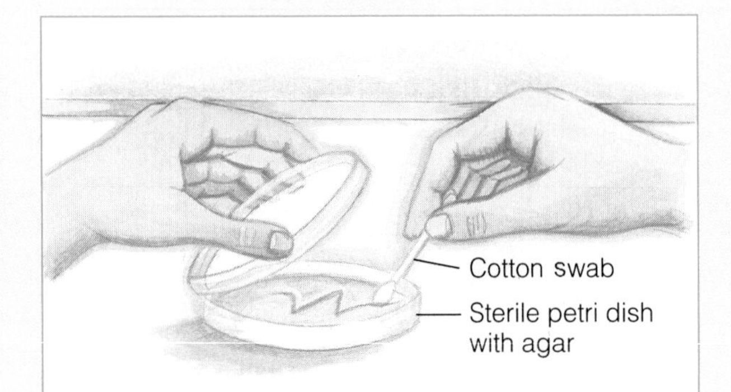

Observations

DATA TABLE

Petri Dish	Day 1	Day 2	Day 3	Day 4
A				
B				

Analysis and Conclusions

1. Based on the results of this activity, what conditions are necessary for the growth of bacteria?
2. What are some things you could do to slow down the growth of bacteria?
3. Share your results with the class. In what locations were bacteria found to be present?

The freezing point of water is 0°C. This is the temperature at which liquid water freezes and becomes solid ice. In the presence of genetically engineered ice-minus bacteria, the freezing point of water can be lowered to -5°C. Is there any other way to lower the freezing point of water? Try this activity to find out.

Materials

- ice cubes
- water
- Styrofoam cup
- salt
- plastic spoon
- Celsius thermometer

Procedure

1. Put some ice cubes into a Styrofoam cup and fill the cup with water.
2. Place a Celsius thermometer in the ice-water mixture. Wait a few minutes and then read the temperature on the thermometer. What is the temperature of the ice-water mixture?

3. Remove the thermometer. Add one or two spoonfuls of salt to the ice-water mixture and stir to dissolve. Replace the thermometer.

4. Wait a few minutes and then read the temperature. What is the temperature of the ice-water mixture plus the dissolved salt? What effect did adding salt to the mixture have on the temperature?

Think for Yourself

Based on your observations, why do you think people often sprinkle rock salt on icy sidewalks?

Appendix A

THE METRIC SYSTEM

The metric system of measurement is used by scientists throughout the world. It is based on units of ten. Each unit is ten times larger or ten times smaller than the next unit. The most commonly used units of the metric system are given below. After you have finished reading about the metric system, try to put it to use. How tall are you in metrics? What is your mass? What is your normal body temperature in degrees Celsius?

Commonly Used Metric Units

Length The distance from one point to another

meter (m)	A meter is slightly longer than a yard. 1 meter = 1000 millimeters (mm) 1 meter = 100 centimeters (cm) 1000 meters = 1 kilometer (km)

Volume The amount of space an object takes up

liter (L)	A liter is slightly more than a quart. 1 liter = 1000 milliliters (mL)

Mass The amount of matter in an object

gram (g)	A gram has a mass equal to about one paper clip. 1000 grams = 1 kilogram (kg)

Temperature The measure of hotness or coldness

degrees Celsius (°C)	0°C = freezing point of water 100°C = boiling point of water

Metric–English Equivalents

2.54 centimeters (cm) = 1 inch (in.)
1 meter (m) = 39.37 inches (in.)
1 kilometer (km) = 0.62 miles (mi)
1 liter (L) = 1.06 quarts (qt)
250 milliliters (mL) = 1 cup (c)
1 kilogram (kg) = 2.2 pounds (lb)
28.3 grams (g) = 1 ounce (oz)
$°C = 5/9 \times (°F - 32)$

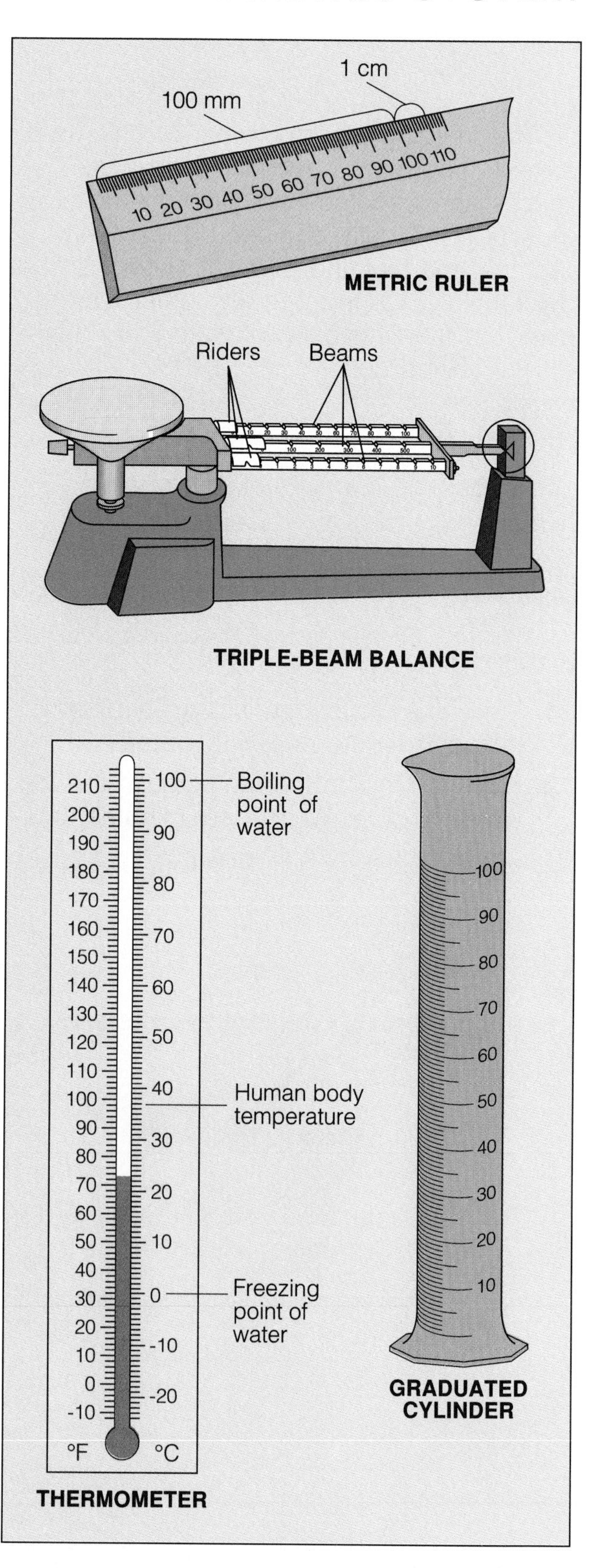

Appendix B

LABORATORY SAFETY
Rules and Symbols

Glassware Safety

1. Whenever you see this symbol, you will know that you are working with glassware that can easily be broken. Take particular care to handle such glassware safely. And never use broken or chipped glassware.
2. Never heat glassware that is not thoroughly dry. Never pick up any glassware unless you are sure it is not hot. If it is hot, use heat-resistant gloves.
3. Always clean glassware thoroughly before putting it away.

Fire Safety

1. Whenever you see this symbol, you will know that you are working with fire. Never use any source of fire without wearing safety goggles.
2. Never heat anything—particularly chemicals—unless instructed to do so.
3. Never heat anything in a closed container.
4. Never reach across a flame.
5. Always use a clamp, tongs, or heat-resistant gloves to handle hot objects.
6. Always maintain a clean work area, particularly when using a flame.

Heat Safety

Whenever you see this symbol, you will know that you should put on heat-resistant gloves to avoid burning your hands.

Chemical Safety

1. Whenever you see this symbol, you will know that you are working with chemicals that could be hazardous.
2. Never smell any chemical directly from its container. Always use your hand to waft some of the odors from the top of the container toward your nose—and only when instructed to do so.
3. Never mix chemicals unless instructed to do so.
4. Never touch or taste any chemical unless instructed to do so.
5. Keep all lids closed when chemicals are not in use. Dispose of all chemicals as instructed by your teacher.
6. Immediately rinse with water any chemicals, particularly acids, that get on your skin and clothes. Then notify your teacher.

Eye and Face Safety

1. Whenever you see this symbol, you will know that you are performing an experiment in which you must take precautions to protect your eyes and face by wearing safety goggles.
2. When you are heating a test tube or bottle, always point it away from you and others. Chemicals can splash or boil out of a heated test tube.

Sharp Instrument Safety

1. Whenever you see this symbol, you will know that you are working with a sharp instrument.
2. Always use single-edged razors; double-edged razors are too dangerous.
3. Handle any sharp instrument with extreme care. Never cut any material toward you; always cut away from you.
4. Immediately notify your teacher if your skin is cut.

Electrical Safety

1. Whenever you see this symbol, you will know that you are using electricity in the laboratory.
2. Never use long extension cords to plug in any electrical device. Do not plug too many appliances into one socket or you may overload the socket and cause a fire.
3. Never touch an electrical appliance or outlet with wet hands.

Animal Safety

1. Whenever you see this symbol, you will know that you are working with live animals.
2. Do not cause pain, discomfort, or injury to an animal.
3. Follow your teacher's directions when handling animals. Wash your hands thoroughly after handling animals or their cages.

Appendix C

SCIENCE SAFETY RULES

One of the first things a scientist learns is that working in the laboratory can be an exciting experience. But the laboratory can also be quite dangerous if proper safety rules are not followed at all times. To prepare yourself for a safe year in the laboratory, read over the following safety rules. Then read them a second time. Make sure you understand each rule. If you do not, ask your teacher to explain any rules you are unsure of.

Dress Code

1. Many materials in the laboratory can cause eye injury. To protect yourself from possible injury, wear safety goggles whenever you are working with chemicals, burners, or any substance that might get into your eyes. Never wear contact lenses in the laboratory.

2. Wear a laboratory apron or coat whenever you are working with chemicals or heated substances.

3. Tie back long hair to keep it away from any chemicals, burners and candles, or other laboratory equipment.

4. Remove or tie back any article of clothing or jewelry that can hang down and touch chemicals and flames.

General Safety Rules

5. Read all directions for an experiment several times. Follow the directions exactly as they are written. If you are in doubt about any part of the experiment, ask your teacher for assistance.

6. Never perform activities that are not authorized by your teacher. Obtain permission before "experimenting" on your own.

7. Never handle any equipment unless you have specific permission.

8. Take extreme care not to spill any material in the laboratory. If a spill occurs, immediately ask your teacher about the proper cleanup procedure. Never simply pour chemicals or other substances into the sink or trash container.

9. Never eat in the laboratory.

10. Wash your hands before and after each experiment.

First Aid

11. Immediately report all accidents, no matter how minor, to your teacher.

12. Learn what to do in case of specific accidents, such as getting acid in your eyes or on your skin. (Rinse acids from your body with lots of water.)

13. Become aware of the location of the first-aid kit. But your teacher should administer any required first aid due to injury. Or your teacher may send you to the school nurse or call a physician.

14. Know where and how to report an accident or fire. Find out the location of the fire extinguisher, phone, and fire alarm. Keep a list of important phone numbers—such as the fire department and the school nurse—near the phone. Immediately report any fires to your teacher.

Heating and Fire Safety

15. Again, never use a heat source, such as a candle or burner, without wearing safety goggles.

16. Never heat a chemical you are not instructed to heat. A chemical that is harmless when cool may be dangerous when heated.

17. Maintain a clean work area and keep all materials away from flames.

18. Never reach across a flame.

19. Make sure you know how to light a Bunsen burner. (Your teacher will demonstrate the proper procedure for lighting a burner.) If the flame leaps out of a burner toward you, immediately turn off the gas. Do not touch the burner. It may be hot. And never leave a lighted burner unattended!

20. When heating a test tube or bottle, always point it away from you and others. Chemicals can splash or boil out of a heated test tube.

21. Never heat a liquid in a closed container. The expanding gases produced may blow the container apart, injuring you or others.

22. Before picking up a container that has been heated, first hold the back of your hand near it. If you can feel the heat on the back of your hand, the container may be too hot to handle. Use a clamp or tongs when handling hot containers.

Using Chemicals Safely

23. Never mix chemicals for the "fun of it." You might produce a dangerous, possibly explosive substance.

24. Never touch, taste, or smell a chemical unless you are instructed by your teacher to do so. Many chemicals are poisonous. If you are instructed to note the fumes in an experiment, gently wave your hand over the opening of a container and direct the fumes toward your nose. Do not inhale the fumes directly from the container.

25. Use only those chemicals needed in the activity. Keep all lids closed when a chemical is not being used. Notify your teacher whenever chemicals are spilled.

26. Dispose of all chemicals as instructed by your teacher. To avoid contamination, never return chemicals to their original containers.

27. Be extra careful when working with acids or bases. Pour such chemicals over the sink, not over your workbench.

28. When diluting an acid, pour the acid into water. Never pour water into an acid.

29. Immediately rinse with water any acids that get on your skin or clothing. Then notify your teacher of any acid spill.

Using Glassware Safely

30. Never force glass tubing into a rubber stopper. A turning motion and lubricant will be helpful when inserting glass tubing into rubber stoppers or rubber tubing. Your teacher will demonstrate the proper way to insert glass tubing.

31. Never heat glassware that is not thoroughly dry. Use a wire screen to protect glassware from any flame.

32. Keep in mind that hot glassware will not appear hot. Never pick up glassware without first checking to see if it is hot. See #22.

33. If you are instructed to cut glass tubing, fire-polish the ends immediately to remove sharp edges.

34. Never use broken or chipped glassware. If glassware breaks, notify your teacher and dispose of the glassware in the proper trash container.

35. Never eat or drink from laboratory glassware. Thoroughly clean glassware before putting it away.

Using Sharp Instruments

36. Handle scalpels or razor blades with extreme care. Never cut material toward you; cut away from you.

37. Immediately notify your teacher if you cut your skin when working in the laboratory.

Animal Safety

38. No experiments that will cause pain, discomfort, or harm to mammals, birds, reptiles, fishes, and amphibians should be done in the classroom or at home.

39. Animals should be handled only if necessary. If an animal is excited or frightened, pregnant, feeding, or with its young, special handling is required.

40. Your teacher will instruct you as to how to handle each animal species that may be brought into the classroom.

41. Clean your hands thoroughly after handling animals or the cage containing animals.

End-of-Experiment Rules

42. After an experiment has been completed, clean up your work area and return all equipment to its proper place.

43. Wash your hands after every experiment.

44. Turn off all burners before leaving the laboratory. Check that the gas line leading to the burner is off as well.

Appendix D

THE MICROSCOPE

The microscope is an essential tool in the study of life science. It enables you to see things that are too small to be seen with the unaided eye. It also allows you to look more closely at the fine details of larger things.

The microscope you will use in your science class is probably similar to the one illustrated on the following page. This is a compound microscope. It is called compound because it has more than one lens. A simple microscope would contain only one lens. The lenses of the compound microscope are the parts that magnify the object being viewed.

Typically, a compound microscope has one lens in the eyepiece, the part you look through. The eyepiece lens usually has a magnification power of 10X. That is, if you were to look through the eyepiece alone, the object you were viewing would appear 10 times larger than it is.

The compound microscope may contain one or two other lenses. These two lenses are called the low- and high-power objective lenses. The low-power objective lens usually has a magnification of 10X. The high-power objective lens usually has a magnification of 40X. To figure out what the total magnification of your microscope is when using the eyepiece and an objective lens, multiply the powers of the lenses you are using. For example, eyepiece magnification (10X) multiplied by low-power objective lens magnification (10X) = 100X total magnification. What is the total magnification of your microscope using the eyepiece and the high-power objective lens?

To use the microscope properly, it is important to learn the name of each part, its function, and its location on your microscope. Keep the following procedures in mind when using the microscope:

1. Always carry the microscope with both hands. One hand should grasp the arm, and the other should support the base.

2. Place the microscope on the table with the arm toward you. The stage should be facing a light source.

3. Raise the body tube by turning the coarse adjustment knob.

4. Revolve the nosepiece so that the low-power objective lens (10X) is directly in line with the body tube. Click it into place. The low-power lens should be directly over the opening in the stage.

5. While looking through the eyepiece, adjust the diaphragm and the mirror so that the greatest amount of light is coming through the opening in the stage.

6. Place the slide to be viewed on the stage. Center the specimen to be viewed over the hole in the stage. Use the stage clips to hold the slide in position.

7. Look at the microscope from the side rather than through the eyepiece. In this way, you can watch as you use the coarse adjustment knob to lower the body tube until the low-power objective almost touches the slide. Do this slowly so you do not break the slide or damage the lens.

8. Now, looking through the eyepiece, observe the specimen. Use the coarse adjustment knob to raise the body tube, thus raising the low-power objective away from the slide. Continue to raise the body tube until the specimen comes into focus.

9. When viewing a specimen, be sure to keep both eyes open. Though this may seem strange at first, it is really much easier on your eyes. Keeping one eye closed may create a strain, and you might get a headache. Also, if you keep both eyes open, it is easier to draw diagrams of what you are observing. In this way, you do not have to turn your head away from the microscope as you draw.

10. To switch to the high-power objective lens (40X), look at the microscope from the side. Now, revolve the nosepiece so that the high-power objective lens clicks into place. Make sure the lens does not hit the slide.

11. Looking through the eyepiece, use only the fine adjustment knob to bring the specimen into focus. Why should you not use the coarse adjustment knob with the high-power objective?

12. Clean the microscope stage and lens when you are finished. To clean the lenses, use lens paper only. Other types of paper may scratch the lenses.

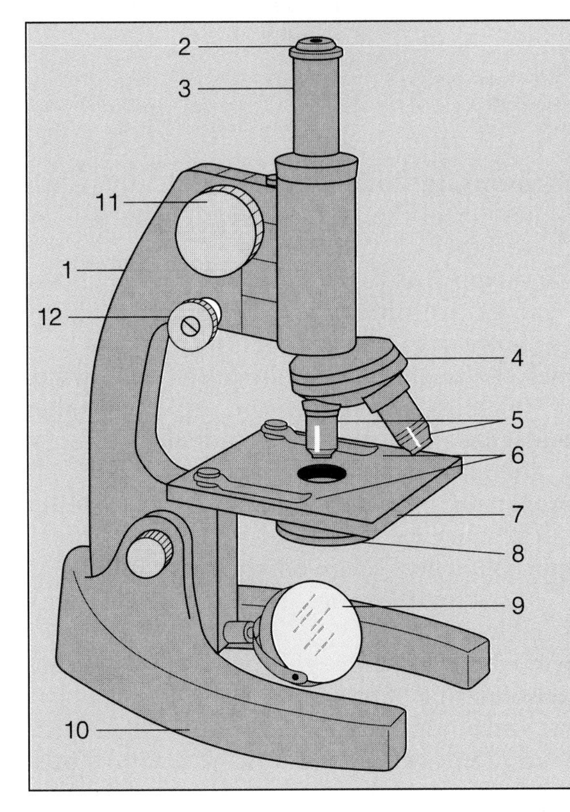

Microscope Parts and Their Functions

1. **Arm** Supports the body tube
2. **Eyepiece** Contains the magnifying lens you look through
3. **Body tube** Maintains the proper distance between the eyepiece and the objective lenses
4. **Nosepiece** Holds the high- and the low-power objective lenses and can be rotated to change magnification
5. **Objective lenses** A low-power lens, which usually provides 10X magnification, and a high-power lens, which usually provides 40X magnification
6. **Stage clips** Hold the slide in place
7. **Stage** Supports the slide being viewed
8. **Diaphragm** Regulates the amount of light let into the body tube
9. **Mirror** Reflects the light upward through the diaphragm, the specimen, and the lenses
10. **Base** Supports the microscope
11. **Coarse adjustment knob** Moves the body tube up and down for focusing
12. **Fine adjustment knob** Moves the body tube slightly to sharpen the image

Glossary

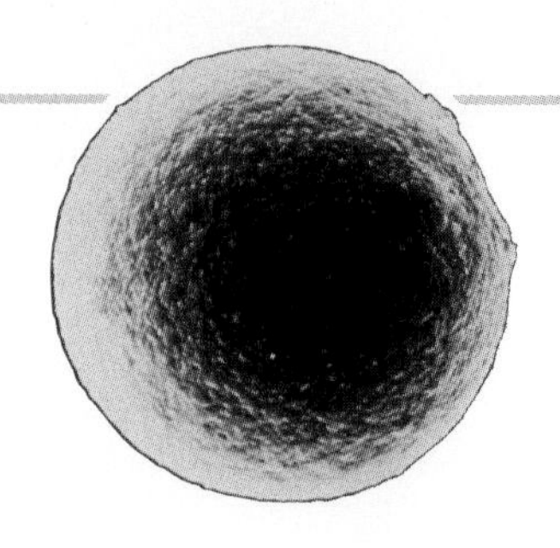

allele (uh-LEEL): each member of a gene pair that determines a specific trait

amino acids: building blocks of proteins

amniocentesis (am-nee-oh-sehn-TEE-sihs): process that involves the removal of a small amount of fluid from the sac that surrounds a developing baby; used to detect genetic disorders

chromosomes: rod-shaped structures found in the nucleus of every cell in an organism

codominant: alleles that are both expressed when both are inherited

deoxyribonucleic (dee-AHKS-ih-righ-boh-noo-KLEE-ihk) **acid:** molecule that stores and passes on genetic information from one generation to the next

DNA: deoxyribonucleic acid

dominant: trait that is expressed when two different genes for the same trait are present; "stronger" of two traits

genes: units of heredity; segments of DNA on chromosomes

genetic engineering: process in which genes, or pieces of DNA, from one organism are transferred into another organism

genetics (juh-NEHT-ihks): study of heredity, or the passing on of traits from an organism to its offspring

genotype (JEHN-uh-tighp): gene makeup of an organism

hybrid (HIGH-brihd): organism that has two different genes for a trait, or that combines traits of two different but related species

hybridization (high-brihd-ih-ZAY-shuhn): crossing of two genetically different but related species of organisms

inbreeding: crossing plants or animals that have the same or similar sets of genes, rather than different genes

incomplete dominance: condition in which neither of the two genes in a gene pair masks the other

karyotype (KAR-ee-uh-tighp): chart that shows the size, number, and shape of all the chromosomes in an organism

meiosis (migh-OH-sihs): process of cell division in which sex cells (sperm and egg) are produced

mutagens: factors, such as radiation and certain chemicals, that cause mutations

mutation: sudden change in a gene or chromosome

nondisjunction (nahn-dihs-JUHNGK-shuhn): failure of a chromosome pair to separate during meiosis

phenotype (FEE-noh-tighp): physical appearance

plasmid: ring of bacterial DNA

recessive: trait that seems to disappear when two different genes for the same trait are present; "weaker" of two traits

recombinant DNA: DNA that contains DNA from two different organisms

replication (rehp-luh-KAY-shuhn): process in which DNA molecules form exact duplicates

ribonucleic acid: nucleic acid that "reads" the genetic information carried by DNA and guides protein synthesis

RNA: ribonucleic acid

selective breeding: crossing of plants and animals that have desirable characteristics to produce offspring with those desirable characteristics

sex chromosomes: chromosomes that determine the sex of an organism; X and Y chromosomes

sex-linked traits: traits that are carried on the X chromosome

traits: physical characteristics

Index

Credits

Cover Background: Ken Karp
Photo Research: Natalie Goldstein
Contributing Artists: Jeanni Brunnick/Christine Prapas, Art Representatives; Warren Budd Assoc. Ltd.; Function Thru Form; Holly Jones/Cornell and McCarthy, Art Representatives; David Biedrzycki; Martinu Schneegass; Gerry Schrenk; Mark Schuller
Photographs: **4** Kjell B. Sandved/Visuals Unlimited; **5** top: Robert Frerck/Odyssey Productions; bottom: Dwight Kuhn Photography; **6** top: Lefever/Grushow/Grant Heilman Photography; center: Index Stock Photography, Inc.; bottom: Rex Joseph; **8** left: Tibor Bognar/Stock Market; right: Ronnie Kaufman/Stock Market; **9** Philippe Plailly/Science Photo Library/Photo Researchers, Inc.; **10** and **11** Ron Kimball Studios; **12** Bettmann Archive; **13** Jane Grushow/Grant Heilman Photography; **17** all photos: Runk/Schoenberger/Grant Heilman Photography; **18** left: Robert Pearcy/Animals Animals/Earth Scenes; center: Chanan Photography; right: Richard Kolar/Animals Animals/Earth Scenes; **20** left: Dwight Kuhn Photography; right: D. Cavagnaro/DRK Photo; **21** left: Stephen J. Krasemann/DRK Photo; right: DPI; **22** left: Larry Johnson; right: J. Cunningham/Visuals Unlimited; **24** left: Barry L. Runk/Grant Heilman Photography; right: Ronnie Kaufman/Stock Market; **32** and **33** Oxford Molecular Biophysics Laboratory/Science Photo Library/Photo Researchers, Inc.; **34** left: Dwight Kuhn/DRK Photo; right: David M. Phillips/Visuals Unlimited; **35** Dr. Tony Brain/Science Photo Library/Photo Researchers, Inc.; **36** Belinda Wright/DRK Photo; **38** E. R. Degginger/Animals Animals/Earth Scenes; **40** left: L. W. Richardson/Photo Researchers, Inc.; right: John Cancalosi/DRK Photo; **41** left: Barry L. Runk/Grant Heilman Photography; right: Runk/Schoenberger/Grant Heilman Photography; **42** top: J. Langevin/Sygma; top inset: Exxon, USA; bottom: Charles West/Stock Market; **43** top: E. R. Degginger/Animals Animals/Earth Scenes; top inset: Richard Walters/Visuals Unlimited; bottom: Nathan Benn/Woodfin Camp & Associates; **44** A. C. Barrington Brown/Cold Spring Harbor Laboratory; **45** Science Source/Photo Researchers, Inc.; **46** Erich Hartmann/Magnum Photos, Inc.; **48** top: John Cancalosi/Peter Arnold, Inc.; bottom: Dan McCoy/Rainbow; **50** Lennart Nilsson, BEHOLD MAN/Little, Brown & Company © Boehringer Ingelheim International GmbH; **55** Ron Kimball Studios; **56** and **57** Pix-Elation/Rainbow; **58** Luis Castañeda/Image Bank; **59** top: Dan McCoy/Rainbow; bottom: © Lennart Nilsson, THE INCREDIBLE MACHINE/National Geographic Society; **60** top: David M. Phillips/Visuals Unlimited; bottom: John Giannicchi/Science Source/Photo Researchers, Inc.; **61** Michel Tcherevkoff/Image Bank; **62** Christine L. Case/Visuals Unlimited; **63** left and right: Stanley Flegler/Visuals Unlimited; **65** left: Roy Morsch/Stock Market; right: Robert E. Daemmrich/Tony Stone Worldwide/Chicago Ltd.; **67** Bryan Reynolds/Sygma; **69** Sonya Jacobs/Stock Market; **71** William McCoy/Rainbow; **72** Will and Deni McIntyre/Photo Researchers, Inc.; **73** Bettmann Archive; **78** and **79** Dr. R. L. Brinster/University of Pennsylvania School of Veterinary Medicine; **80** left: David W. Hamilton/Image Bank; right: John Colwell/Grant Heilman Photography; **81** left: R. Van Nostrand/Photo Researchers, Inc.; right: George D. Lepp/Bio-Tec Images; **82** left: Dahlgren/Stock Market; center: Jeff Lepore/Photo Researchers, Inc.; right: Wayne Lankinen/Bruce Coleman, Inc.; **83** Ron Kimball Studios; **84** top: Keith V. Wood/UC San Diego; center: Jon Gordon/Phototake; bottom: Ted Spiegel/Black Star; **85** P. A. McTurk, Univ. of Leicester & D. Parker/Science Photo Library/Photo Researchers, Inc.; **86** top: David M. Phillips/Visuals Unlimited; bottom: Dr. Jeremy Burgess/Science Photo Library/Photo Researchers, Inc.; **87** top: Myrleen Ferguson/Photoedit; center: Omikron/Science Source/Photo Researchers, Inc.; bottom: Norm Thomas/Photo Researchers, Inc.; **88** Jim Tuten/Black Star; **89** Alabama Agricultural Experiment Station/Auburn University; **94** Nik Kleinberg/Picture Group; **98** Steve Maslowski/Photo Researchers, Inc.; **100** Jon Feingersh/Stock Market; **102** John Cancalosi/DRK Photo; **120** John Giannicchi/Science Source/Photo Researchers, Inc.; **121** George D. Lepp/Bio-Tec Images